T0251582

Amarjit S. Basra, PhD

Plant Growth Regulators in Agriculture and Horticulture
Their Role and Commercial Uses

Pre-publication
REVIEWS,
COMMENTARIES,
EVALUATIONS . . .

"To those working in the field of plant growth regulation, a book on the role and commercial uses of PGRs is highly welcome, since only very few treatises of this type have been published in recent times. *Plant Growth Regulators in Agriculture and Horticulture* is particularly valuable to the experienced scientist, since it contains many stimulating ideas and suggestions.

In the nine chapters of the book, emphasis is placed on PGR uses in horticultural crops. Detailed advice on how to use different types of retar-dants in ornamentals is given in Chapter 4. Chapter 5 gives an excellent survey on the role of growth regulators in the postharvest behavior of ornaments. Chapters 6 through 9 deal with various points of PGR applications in different types of fruit. Practical and theoretical aspects of manipulating yield formation in cereals are very thoroughly dealt with in Chapter 2. Regulation of root formation and the benefits of reduced gibberellin levels in stress situations are the subjects of Chapters 1 and 3, respectively. Information on general aspects of phytohormones and the underlying biochemical and physiological modes of action of PGRs is given in the different chapters."

Dr. Wilhelm Rademacher
BASI Agricultural Center,
Limburgerhof, Germany

More pre-publication
REVIEWS, COMMENTARIES, EVALUATIONS . . .

"*P* lant Growth Regulators in Agriculture and Horticulture offers new benefits to update readers on the latest developments in roles and commercial uses. One of the major subjects in this text that scientists and chemical companies are continually looking for is compounds that will improve fruit quality without side effects to the environment or to human health and safety. This book puts forth the strategies of crop load management of bloom thinning in pome and stone fruits. Also, new cultivars of fruits are continually being developed whose cultivation and response to PGRs must be tested.

This book provides a sound basis for understanding the use of PGRs in the twenty-first century. It is an excellent overview for growers, breeders, researchers, chemical companies, and students."

Mahamoud T. Nawar, PhD
Chief of Research,
Cotton Research Institute,
Agricultural Resource Centre,
Giza, Egypt

Food Products Press®
An Imprint of The Haworth Press, Inc.
New York • London • Oxford

Plant Growth Regulators in Agriculture and Horticulture

Their Role and Commercial Uses

FOOD PRODUCTS PRESS
Crop Science
Amarjit S. Basra, PhD
Senior Editor

New, Recent, and Forthcoming Titles of Related Interest:

Dictionary of Plant Genetics and Molecular Biology by Gurbachan S. Miglani

Advances in Hemp Research by Paolo Ranalli

Wheat: Ecology and Physiology of Yield Determination by Emilio H. Satorre and Gustavo A. Slafer

Mineral Nutrition of Crops: Fundamental Mechanisms and Implications by Zdenko Rengel

Conservation Tillage in U.S. Agriculture: Environmental, Economic, and Policy Issues by Noel D. Uri

Cotton Fibers: Developmental Biology, Quality Improvement, and Textile Processing edited by Amarjit S. Basra

Heterosis and Hybrid Seed Production in Agronomic Crops edited by Amarjit S. Basra

Intensive Cropping: Efficient Use of Water, Nutrients, and Tillage by S. S. Prihar, P. R. Gajri, D. K. Benbi, and V. K. Arora

Physiological Bases for Maize Improvement edited by María E. Otegui and Gustavo A. Slafer

Plant Growth Regulators in Agriculture and Horticulture: Their Role and Commercial Uses edited by Amarjit S. Basra

Crop Responses and Adaptations to Temperature Stress edited by Amarjit S. Basra

Plant Growth Regulators in Agriculture and Horticulture
Their Role and Commercial Uses

Amarjit S. Basra, PhD
Editor

Published by

Food Products Press®, an imprint of The Haworth Press, Inc., 10 Alice Street, Binghamton, NY 13904-1580

Cover design by Monica L. Seifert.

Library of Congress Cataloging-in-Publication Data

Plant growth regulators in agriculture and horticulture : their role and commercial uses / Amarjit S. Basra, editor.
    p. cm.
   Includes bibliographical references and index.
   ISBN 1-56022-891-1 (hardcover : alk. paper) — ISBN 1-56022-896-2 (softcover : alk. paper)
1. Plant regulators. I. Basra, Amarjit S.

SB128 .P56 2000
631.8'9--dc21

                                      00-039357

CONTENTS

ABOUT THE EDITOR

Amarjit S. Basra, PhD, is an eminent plant physiologist at Punjab Agricultural University in Ludhiana, India. His outstanding research in the field of plant growth regulation has been published in the world's leading journals and is widely cited.

He has over 80 research papers and eight edited books to his credit. He is a member of the American Association of Plant Physiologists, the Australian Society of Plant Physiologists, the Crop Science Society of America, the American Society of Agronomy, the American Institute of Biological Sciences, the American Society of Horticultural Science, and the International Society of Horticultural Sciences.

Dr. Basra has received coveted scientific awards and honors for his outstanding contributions, including the INSA Medal for Young Scientists and the Rafi Ahmad Memorial Prize for Agricultural Research.

He provides leadership in organizing and fostering collaboration in international crop science research and dissemination of information.

CONTRIBUTORS

R. Austin Fletcher, BSc (Hon), MSc, PhD, is Professor Emeritus, Department of Environmental Biology, University of Guelph, Guelph, Ontario, Canada.

Hector E. Flores, PhD, is Professor of Plant Pathology and Biotechnology, Department of Plant Pathology, Pennsylvania State University, University Park, Pennsylvania.

Shmuel Gazit, PhD, is Professor, Kennedy-Leigh Center for Horticultural Research, Hebrew University of Jerusalem, Rehovolt, Israel.

Martin P. N. Gent, PhD, is Associate Scientist, Department of Forestry and Horticulture, Connecticut Agricultural Experiment Station, New Haven, Connecticut.

Kiyohide Kojima, PhD, is Professor, Faculty of Agriculture, Niigata University, Niigata, Japan.

Susan Lurie, PhD, is Professor and Head, Deciduous Fruits Postharvest Unit, Volcani Center, Bet Dagan, Israel.

Richard J. McAvoy, PhD, is Associate Professor, University of Connecticut, Storrs, Connecticut.

Renae E. Moran, PhD, is Assistant Professor, Department of Biosystems Science and Engineering, University of Maine, Monmouth, Maine.

Pirjo Peltonen-Sainio, PhD, is Professor, Agricultural Research Center of Finland, Plant Production Research, Crop and Soil, Jokioinen, Finland.

Ari Rajala, MSc, is Research Scientist, Agricultural Research Center of Finland, Plant Production Research, Crop and Soil, Jokioinen, Finland.

Michael S. Reid, DSc, is Professor, Department of Environmental Horticulture, University of California, Davis, California.

Margrethe Serek, PhD, is Professor, Department of Horticultural Sciences, Royal Veterinary and Agricultural University, Frederiksberg C., Denmark.

Coralie R. Sopher, MSc, is a doctoral candidate, Department of Environmental Biology, University of Guelph, Guelph, Ontario, Canada.

Stephen M. Southwick, PhD, is Pomologist, University of California, Davis, California.

Raphael A. Stern, PhD, is Tree Fruit Physiologist, MIGAL, Galilee Technology Center, Rosha Pina, Israel.

Nataraj N. Vettakkorumakankav, BSc, MSc, PhD, is Research Associate, Department of Environmental Biology, University of Guelph, Guelph, Ontario, Canada.

Jorge M. Vivanco, PhD, is Assistant Professor of Horticultural Biotechnology, Department of Horticulture and Landscape Architecture, Colorado State University, Fort Collins, Colorado.

Preface

Plant growth regulators (PGRs) are organic compounds other than nutrients (supplying either energy or mineral elements) that, in small amounts, promote, inhibit, or otherwise modify any physiological process in plants. The term PGR includes both naturally occurring plant growth substances, or phytohormones, as well as synthetic compounds, or chemical analogs.

There are five well-established categories of "classical" phytohormones, namely, auxins, gibberellins, cytokinins, abscisic acid, and ethylene. More recently, several other compounds that can regulate various facets of plant growth and development have been described, such as oligosaccharins, brassinosteroids, jasmonates, salicylates, and polyamines. The list is likely to grow, with new PGRs waiting to be discovered. The availability of sophisticated methods for the identification and quantitative measurements of PGRs, hormone mutants, and powerful tools of molecular biology has greatly enhanced our understanding of their regulatory role in plant growth and development.

Since the 1940s, both natural and synthetic growth regulators have been used in agriculture and horticulture with increasing incidence to modify crop plants by controlling plant developmental processes, from germination through vegetative growth, reproductive development, maturity, senescence (or aging), and postharvest preservation. In the category of synthetic bioregulators, the primary focus has been on herbicides or fungicides. Estimates of the complexity and the costs associated with the discovery of new PGRs are many times greater than for herbicides. Hence, the weed-killing aspects of plant growth regulation have overshadowed the other uses of PGRs in crop production.

In view of growing environmental concerns, the investment in research and development and the registration of new PGRs has faced difficulties, restricting the vast potential of the effective use of PGRs for enhancing production efficiency of the world's major agricultural and horticultural crops. A detailed understanding of their physiological roles and modes of action will not only aid the discovery and synthesis of new PGR products but also will assist in predicting and preventing their secondary effects. It is, therefore, very timely to make a critical appraisal of this field, dealing with both basic and applied aspects.

The aim of this book is to provide a well-founded evaluation of the commercial uses of PGRs in agriculture and horticulture, using selected examples, along with rapidly accelerating knowledge of their regulatory roles in plant growth and development, employing modern experimental methods. This integrated approach makes the book useful for a wide range of professionals, especially those in the areas of plant physiology, agricultural chemistry, agronomy, soil science, horticulture, forestry, environmental biology, and many other related disciplines. This book attempts to synthesize and filter a vast amount of primary and applied research data in the field of plant growth regulation. It also provides comprehensive reference lists for further exploration of topics and issues, as well as innovative ideas for future research and development.

This book contains chapters written by highly competent scientists who have pooled their specialized knowledge to create a valuable resource of detailed and credible information for the benefit of the international community. I am grateful to each one of them for making this book a reality.

Bill Cohen and Bill Palmer of The Haworth Press have enthusiastically supported the idea of this publication and provided much encouragement. Andy Roy provided invaluable assistance during the preproduction stages of the book's publication. Peg Marr and Amy Rentner have done a meticulous job of copyediting, adding tremendous value to the final product. Melissa Devendorf was always there, supportive and ever efficient.

Finally, what a pleasure to acknowledge the loving support and inspiration of my wife, Ranjit; daughter, Sukhmani; and son, Nishchayjit, which kept me going!

Amarjit Singh Basra

Chapter 1

Control of Root Formation by Plant Growth Regulators

Jorge M. Vivanco
Hector E. Flores

INTRODUCTION AND HISTORICAL PERSPECTIVE

Almost 100 years have passed since F. W. Went's pioneering work on plant growth regulators, or phytohormones. At that time, the term phytohormones referred almost entirely to auxins, although it was suspected that other phytohormones existed, based on physiological experiments (Went and Thimann, 1937). It was not until 1946 that Haagen-Smit and others isolated pure indole-3-acetic acid (IAA) from the endosperm of immature corn grains, although IAA was previously isolated and described in fermentation media (Salkowski, 1885), human urine (Kogl and Haagen-Smit, 1931), yeast (Kogl and Kostermans, 1934), and in *Rhizopus suinus* (Thimann, 1935b).

In 1935, Yabuta isolated an active crystalline material from *Gibberella fujikoroi,* also known as the bakanae (foolish seedling) fungus on rice. This substance was found to stimulate growth when applied to the roots of rice seedlings and was called gibberellin A. Since 1937, gibberellins, cytokinins, ethylene, and abscisic acid have joined auxins as plant growth regulators, or phytohormones, and are known as the "classical five."

More recently, several other compounds that affect plant growth and development have been described. In many cases, the roles that these compounds play in modulating growth and development have been defined by the analysis of mutants and transgenic plants, with either altered perception or levels of the phytohormones under study. Among the "nontraditional" plant growth regulators are oligosaccharins, brassinosteroids, jasmonates, salicylic acid, and polyamines (Flores, Young, and Galston, 1984; Raskin, 1992; Aldington and Fry, 1993; Sakurai and Fujioka, 1994; Arteca, 1996). Far less is known about compounds such as turgorins, strigols, and other promotors/inhibitors of seed germination and plant growth (Gross and Parthier, 1994). Moreover, the number of "nontraditional" phytohormones is expected to grow as new compounds with growth-regulating properties are discovered.

Most of the current information available on plant growth regulators is associated with whole-plant physiology responses. However, the specific regulatory roles of these substances in root formation and development are beginning to be elucidated. New approaches, such as molecular genetics, will help to unravel aspects of hormone root physiology. In this chapter, we will describe the regulatory behavior of the classical five and some nontraditional plant growth regulators in the control of root formation, with emphasis on adventitious and "hairy root" formation. Also, we will discuss the *Agrobacterium rhizogenes* hairy root hormonal-formation mechanism.

PLANT GROWTH REGULATORS INVOLVED IN ROOT FORMATION AND DEVELOPMENT

In 1976, evidence of the presence of the five major plant hormones was suggested, but the data remained inconclusive as to whether plant growth regulators were synthesized in the roots (Torrey, 1976). Recently, it has been suggested that jasmonic acid should be added to those five plant hormones, since evidence of its presence in roots has been established (Gross and Parthier, 1994). In this section, the biochemistry, biosynthesis, metabolism, and regulatory mechanisms associated with root formation and development of the classical five plant hormones (see Figure 1.1) and some other nontraditional hormones (see Figure 1.2) will be examined. Moreover, evidence that these plant growth regulators are produced in roots will be explored.

FIGURE 1.1. Structures of the "Classical Five" Plant Hormones

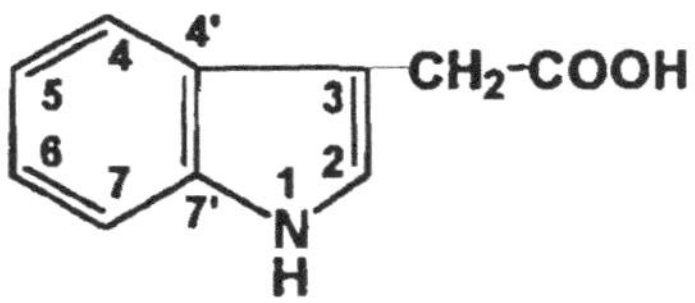

Indole-3-acetic acid

$H_2C{=}CH_2$

Ethylene

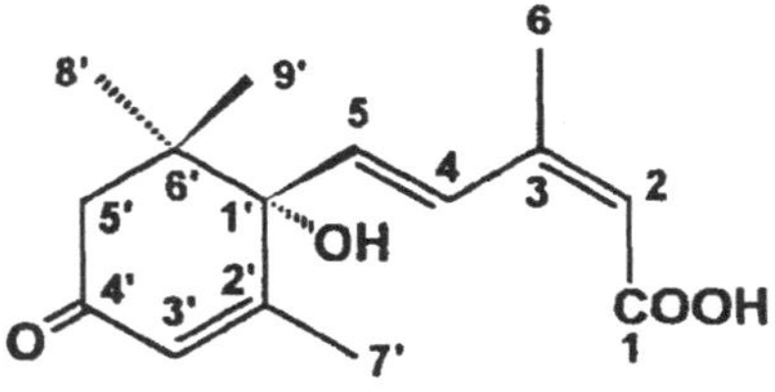

***S*-(+)-Abscisic acid**

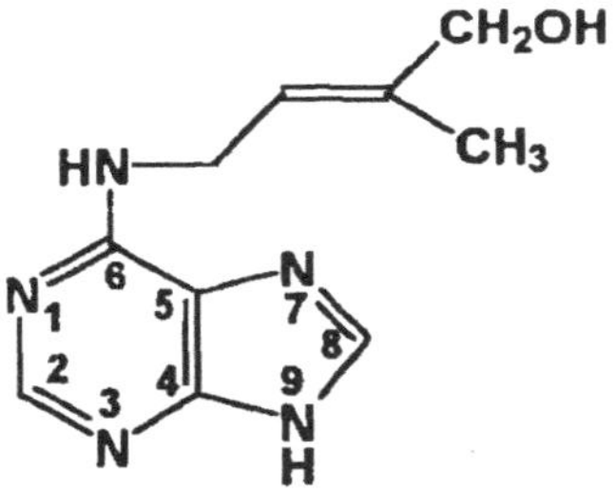

Zeatin

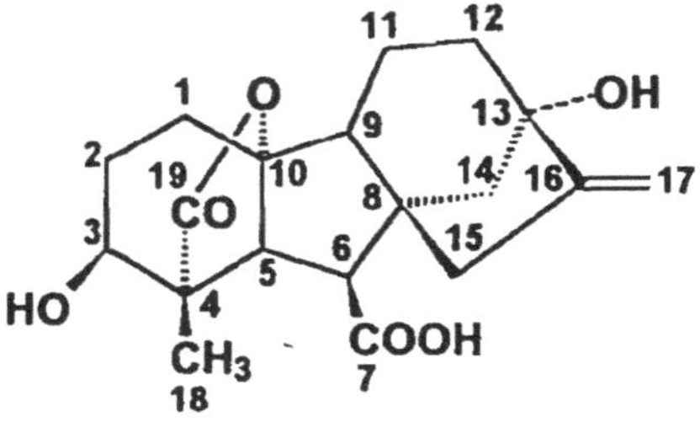

Gibberellin A_1

FIGURE 1.2. Structures of the "Nontraditional" Plant Hormones

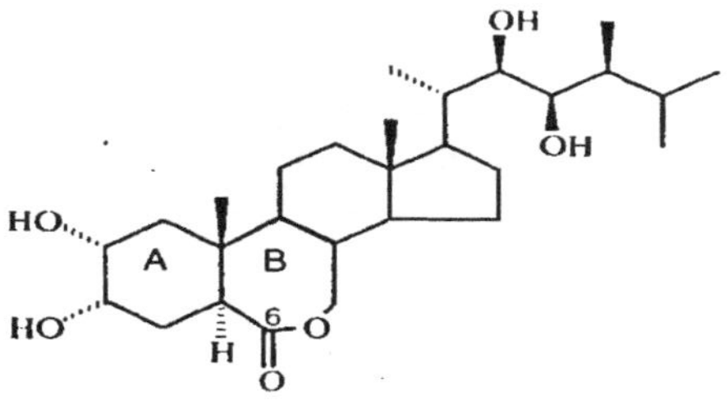

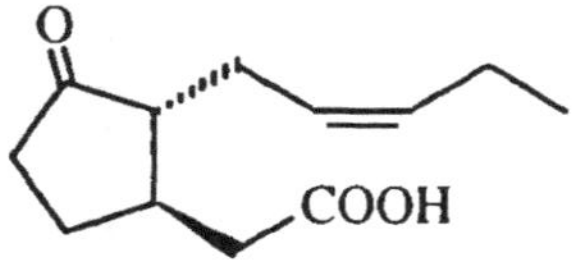

Jasmonic acid

Brassinolide

$H_2N-(CH_2)_4-NH_2$

Putrescine

$H_2N-(CH_2)_3-NH-(CH_2)_4-NH-(CH_2)_3-NH_2$

Spermine

$H_2N-(CH_2)_4-NH-(CH_2)_3-NH_2$

Spermidine

Auxins

The primary auxin in plants is IAA. Following the discovery of IAA, many indole compounds have been isolated in plants, such as indole-3-butyric acid, phenylacetic acid and 4-chloro-IAA (Normanly, Slovin, and Cohen, 1995). However, auxin activity of these compounds could be attributed to their conversion to IAA (Arteca, 1996). For many years, it was assumed that tryptophan was the precursor of IAA. This has been confirmed biochemically, utilizing different model systems (Wildman, Ferri, and Bonner, 1947; Muir and Lantican, 1968), and has recently been reconfirmed in seedlings of *Phaseolus vulgaris,* using radioisotope studies (Bialek, Michalczuk, and Cohen, 1992). Three routes for IAA biosynthesis from tryptophan have been proposed, via indole-3-pyruvic acid, tryptamine, or indole-3-acetonitrile (Cohen and Bialek, 1984).

Recently, work with tryptophan auxotrophic mutants has shown that IAA biosynthesis can also take place by a tryptophan independent route. It was found that Arabidopsis tryptophan auxotrophs accumulate more IAA than do wild-type plants. Based on this information, a pathway was proposed, in which IAA is synthesized through a branch point of the tryptophan biosynthetic pathway at indole or indole-glycerol phosphate (Normanlyal, Slovin, and Cohen, 1995). Certain bacteria and plant cells transformed with *Agrobacterium tumefaciens* also synthesize IAA via a unique pathway, in which tryptophan is converted to indole-3-acetamide, which in turn is converted to IAA (Klee and Romano, 1994). Conversely, the catabolism of auxins is basically due to oxidation to oxindole-3-acetic acid and subsequent glycosylation (Normanly, Slovin, and Cohen, 1995). Another catabolic pathway is via IAA-acetylaspartate to dioxindole-3-acetylaspartate-3-O-glucoside.

Two types of auxin are found in plants: free and bound. Free auxins can rapidly diffuse out of the tissue and, thus, be used immediately to regulate physiological processes. On the other hand, bound auxins are usually conjugated to amino acids, peptides, or carbohydrates and are made available only after they are subjected to hydrolysis, enzymolysis, or autolysis. Bound auxins typically serve as reserve or storage (glucosides) and detoxification (amino acid or protein complexes) forms of auxin (Cohen and Bialek, 1984).

Research with isolated maize roots indicated that IAA was synthesized in excised roots (Elliott, 1977). Later, it was discovered that all parts of the root were able to synthesize IAA. However, either the root cap or the quiescent center constitutes the place where IAA is most actively metabolized (Feldman, 1980). The physiological roles of auxins have many implications in plant growth, such as cellular elongation, phototropism, geotropism, apical dominance, root initiation, ethylene production, and fruit development. However, this chapter focuses on root-related responses.

Several stages of root formation exhibit extreme sensitivity to exogenous auxin and are correlated in shifts in endogenous auxin concentration (King et al., 1995). Studies performed with a recessive Arabidopsis nuclear mutant, rooty (*rty*), which shows extreme proliferation of roots and inhibition of shoot growth, have elucidated some of the roles of auxin. Apparently, auxin is highly involved in root proliferation and restriction of shoot growth (King et al., 1995).

Cytokinins

Cytokinins occurring in plants are N^6-substituted adenine derivatives. Several bacteria, including *Agrobacterium*, produce cytokinins (Gaudin, Vrain, and Jouanin, 1994). It is known that the initial steps of the mevalonic acid pathway up to the isopentenyl pyrophosphate step are involved in the production of cytokinins. Essentially, isopentenyl pyrophosphate together with adenosine monophosphate (AMP) produces isopentenyl AMP, which is converted to isopentenyladenosine, followed by a series of steps to produce cytokinins (Kaminek, Mok, and Zazimalova, 1992). Isopentenyladenosine 5'-monophosphate is the precursor of all other forms of cytokinins. The ribotides of zeatin and dihydrozeatin are formed through hydroxylation of the isopentenyl side chain and reduction of the double bond (Martin, Mok, and Mok, 1993).

Two major forms of cytokinins exist in plants, free and bound. Examples of free cytokinins are zeatin and isopentenyladenine. The conjugated cytokinins can be produced in different ways, such as glucosides or alanine conjugates (Arteca, 1996). The glucoside conjugates may function as storage forms or, in some cases, may facilitate transport of certain cytokinins, whereas alanine conjugates are irreversibly formed products that may play a role in detoxification mechanisms in the plant. Cytokinins are inactivated by two different reactions: formation of N-conjugates with glucose or with alanine, or by the oxidative cleavage of the N^6 side chain of the cytokinin substrate by cytokinin oxidase (Motyka et al., 1996).

Strong evidence suggests that cytokinins are biosynthesized in the roots and subsequently transported via the xylem to the shoots, where they exert a major regulatory influence on growth, photosynthesis, and timing of senescence (Itai and Birnbaum, 1996). The presence of cytokinins in root extracts and in root exudates supports the view that roots are a site for cytokinin biosynthesis, and that riboside is the main translocated form (Letham and Palni, 1983). It has been established in pea (*Pisum sativum* L.) that the meristematic cells around the quiescent center at the root tip, as well as the root cap, are the major sites of cytokinin biosynthesis in the roots (Zavala and Brandon, 1983). However, in the roots of carrot (*Daucus carota* L.), the conversion of [8]adenine to cytokinin depends on the presence of active cambium tissue (Chen et al., 1985).

The physiological role of cytokinins in root initiation and development is quite ambiguous. Depending on a particular plant and concentration, cytokinins can either promote and/or inhibit root initiation and development. For instance, it has been shown that kinetin (a type of cytokinin) can stimulate dry weight and elongation of roots in lupin seedlings, whereas at higher concentrations both are inhibited (Fries, 1960). Also, the elongation of roots of wheat, flax, and cucumber seedlings was strongly inhibited by various native and synthetic kinetins (Stendil, 1982). Kinetin applied to roots at very low concentrations was able to stimulate photosynthesis and growth. However, growth of the roots and the entire plant was severely diminished if roots were in contact with kinetin for longer than two days (Dong and Arteca, 1982).

Gibberellins

More than 112 gibberellins (GAs) have been identified to date (Hisamatsu et al., 1998). Gibberellins are terpenoids and, thus, are built from five carbon isoprene units. The intermediate precursor of gibberellins is a diterpene that contains four isoprene units (Arteca, 1996). The mevalonic acid pathway is used for the biosynthesis of gibberellins. Each of the steps in this pathway up to GA_{12} aldehyde are the same in all plants. However, from this point on, different species use unique pathways to form the distinct gibberellins (Kende and Zeevaart, 1997). Only a few GAs are bioactive, whereas the others are precursors or deactivated GAs. A 3ß-hydroxyl group (as in GA_1

and GA_4) is required for activity, as was demonstrated with the dwarf *le* mutant of pea (Ingram et al., 1984). Bound gibberellins exist as glucosides, but their role is still not clear (Ross et al., 1995).

Young leaves constitute the major site of biosynthesis of gibberellin, which can subsequently be transported throughout the plant in a nonpolar fashion. However, roots have been demonstrated to be a site of interconversion of gibberellins produced in the shoots rather than a primary source (Crozier and Reid, 1971; Torrey, 1976; Metzger and Zeevaart, 1980). According to Crozier and Reid (1971), GA_{19} is produced in the shoots of *Phaseolus coccinensis* and is subsequently translocated to the roots. In the roots, GA_{19} is converted to GA_1, which is in turn exported back to the shoots. Seeds also produce gibberellins but are unable to transport them (Takahashi, Phinney, and MacMillan, 1991).

Gibberellins have been shown to regulate starch accumulation and utilization inside and outside the root, thus coordinating overall plant growth (Ben-Gad, Altman, and Monselise, 1979). Apparently, gibberellins are able to stimulate the expression of hydrolytic enzymes, such as amylases, which promote the starch to sugar conversion (Brown and Ho, 1986; Davies and Jones, 1991).

Abscisic Acid

Abscisic acid (ABA) is a sesquiterpene composed of three isoprene units. Early on, the similarity in structure between ABA and the end groups of certain carotenoids led to the belief that ABA may be a breakdown product of the carotenoids, with xanthoxin as an intermediate. Today, however, it is known that ABA is synthesized through the early steps of the mevalonic acid pathway, with two potential routes from isopentenyl pyrophosphate (Arteca, 1996; Kende and Zeevaart, 1997). These two routes are via farnesyl pyrophosphate to ABA or carotenoids followed by a series of steps to ABA. In contrast, ABA can be metabolized by conversion to abscisyl-ß-D-glucopuranoside, which is a reversible reaction, or it can be irreversibly converted to 6'-hydroxymethyl ABA, phaseic acid, or 4'-dihydrophaseic acid. ABA can also be inactivated by the formation of an ABA-glucose ester, or due to conjugation of ABA in a similar fashion as to what occurs with auxins, cytokinins, and gibberellins.

Foregoing evidence exists of the presence of ABA in roots and root extracts (Tietz, 1971; Rivier, Milon, and Pilet, 1977). Also, it has been found that aseptic root cultures of *Phaseolus coccinensis,* which never had contact with aerial parts, produced ABA at levels similar to those in intact roots (Hartung and Abou-Mandour, 1980). As with cytokinins, the ABA content of roots is affected by the root environment. Apparently, roots respond to water stress by increasing their ABA levels more quickly and with greater sensitivity than leaves (Cornish and Zeevaart, 1985).

ABA, as well as other inhibiting substances, has been classically postulated to play a role in root growth and gravireaction due to its asymmetric distribution (Feldman, 1982). More recently, it has been shown that ABA is a stimulator of root growth (Abou-Mandour and Hartung, 1980). According to Saab and colleagues (1990), roots are able to maintain their growth under drying soil conditions as long as they are supplied with high amounts of ABA. Clearly, the increased number of root hairs and the increased diameter of ABA-treated roots facilitate the penetration of compacted soil layers by root tips (Eshel and Waisel, 1997; Hartung and Turner, 1997).

Ethylene

Ethylene is a gaseous plant growth substance that has been shown to be involved in numerous aspects of plant growth and development. The immediate precursor of ethylene is 1-aminocyclopropane-1-carboxylic acid (ACC) (Yang and Hoffman, 1984). The key regulatory step in ethylene biosynthesis is the conversion of S-adenosyl-L-methionine to 5'-methyladenosine and ACC. The enzyme that catalyzes this reaction is ACC synthase (Kende, 1993). ACC synthase is encoded by a multigene family whose members are differentially expressed in response to developmental, environmental, and hormonal factors (Olson et al., 1991; Yip, Moore, and Yang, 1992). The final step in ethylene biosynthesis, the conversion of ACC to ethylene, is catalyzed by ACC oxidase (Hamilton, Lycett, and Grierson, 1990). ACC oxidase is also encoded by small multigene families. Although the initial evidence showed that ethylene biosynthesis is controlled by ACC synthase, now it appears that ACC oxidase plays a predominant role in controlling ethylene biosynthesis. Moreover, ethylene biosynthesis comprises a feedback reaction, by which ethylene is able to enhance the expression of ACC synthase and ACC oxidase (Kende, 1993).

Besides being converted to ethylene, ACC can also be irreversibly conjugated to form N-malonyl-ACC (Kionka and Amrhein, 1984). Malonylation of ACC regulates the level of ACC and, thus, the production of ethylene. Furthermore, ethylene can be metabolized by plant tissues to ethylene oxide and ethylene glycol (Sanders, Smith, and Hall, 1989). The physiological role of this process remains unknown. Since ethylene is a gas, it can readily be diffused out of the tissues; thus, no particular metabolism is essential for its removal.

All plant tissues are capable of producing ethylene, which cannot be transported from organ to organ in significant quantities due to its gaseous nature (Yang and Hoffman, 1984). However, ACC (ethylene precursor) occurs in root exudates, and its synthesis is enhanced under anaerobic conditions. As the conversion of ACC to ethylene occurs only under aerobic conditions, ACC was suggested to be the signal that is transported from the roots to the shoots (Bradford and Yang, 1980).

At present, the role of ethylene in root formation and development is not completely understood. However, some indications of its activity exist, such as its role in adventitious root formation (Zimmerman and Hitchcock, 1933; Wample, 1979). Contrarily, it has been reported that ethylene has an inhibitory effect on root growth (Roddick and Guan, 1991). Ethylene elicitation of root exudates has also been shown (Abeles, Morgan, and Saltveit, 1992).

Brassinosteroids

Brassinosteroids (BRs) are a group of naturally occurring polyhydroxy steroids. All known BRs are derivatives of 5-cholestane. Currently, more than sixty BRs have been identified in many plants, including dicots, monocots, gymnosperms, green algae, and ferns (Sakurai and Fujioka, 1994). BRs have been isolated from seeds, fruits, shoots, leaves, and flower buds. BR levels are relatively high in pollen (5-190 nanograms per gram [$ng \cdot g^{-1}$] fresh weight). Brassinolide was the first brassinosteroid characterized and was originally isolated from rape (*Brassica napus* L.) pollen (Grove et al., 1979). At the present time, the biosynthesis of brassinolide has not yet been thoroughly studied. However, it is known that the pathway from capestanol to brassinolide may occur via an early oxidation of campesteronol to form cathasterone (Yokota, 1997).

BRs are powerful inhibitors of root growth and development. BR and IAA effects are generally similar and synergistic in overall plant growth. However, in the case of root initiation, their effects are clearly different (Arteca, 1996). Possibly, the inhibition of root growth is due to BR-induced ethylene production (Roddick and Guan, 1991).

Jasmonates

Jasmonic acid (JA) and its methyl ester, methyl jasmonate (MeJA) are linolenic acid (LA)-derived, cyclopentanone-based compounds. Early studies showed that exogenous JA or MeJA can promote senescence and regulate growth (see Arteca, 1996). Subsequent studies have revealed that JA specifically alters gene expression and that wounding and elicitors can cause JA and MeJA accumulation in plants (Sembdner and Parthier, 1993). It is known that MeJA can elicit physiological responses in a gaseous state (Arteca, 1996). However, transport, intracellular location, and regulation of jasmonates are not well understood.

Jasmonates have been attributed several, though sometimes contradictory, roles in plant growth (van den Berg and Ewing, 1991; Koda, 1992; Parthier et al., 1992; Sembdner and Parthier, 1993). In this regard, exogenous applications of JA strongly inhibit root growth by a mechanism that is not mediated by ethylene (Berger, Bell, and Mullet, 1996). On the other hand, JA induces adventitious root formation (Sembdner and Parthier, 1993). Also, JA or a derivative, tuberonic acid, has been proposed to play a role in the formation of tubers, a specialized vegetative sink (Pelacho and Mingo-Castel, 1991; Koda, 1992; Ravnikar, Vilhar, and Gogala, 1992). Although JA is considered by some to be a new class of plant hormones (Staswick, 1992), we still have major gaps in our understanding and validation of their physiological role as plant hormones.

Polyamines

The diamine putrescine, the triamine spermidine, and the tetramine spermine are ubiquitous in plant cells, while other polyamines are of more limited occurrence. Their chemistry and pathways of biosynthesis and metabolism are well characterized (Flores, Young, and Galston, 1984; Flores, Protacio, and Signs, 1989). They occur in the free form as cations but are often conjugated to small molecules, such as phenolic acids, and also to various macromolecules, such as deoxyribonucleic acid (DNA), ribonucleic acid (RNA), phospholipids, or certain proteins (Cohen, 1971).

Putrescine, spermidine, and spermine are involved in numerous processes associated with plant growth by affecting cell division (Bagni, Serafini-Frascassini, and Torrigiani, 1982) and cell development (Slocum, Kaur-Sawhney, and Galston, 1984). In explants treated with auxin, polyamines show an increase in concentration, especially before root emergence (Desai and Mehta, 1984), but the relationship between auxin-induced root formation, polyamine content, polyamine synthesis inhibitors, and ethylene is still unclear (Biondi et al., 1990). In some species, such as olive, azalea, hazelnut, and poplar, polyamines induce rooting (Rugini and Wang, 1986; Mirkovic, 1993; Rugini et al., 1997). However, it seems likely that polyamines act at the very early stages of the rooting phase in olive. According to Rugini and colleagues (1997), this mechanism is due, not to a higher rate of cell division, but to earlier, root initial formation. Moreover, it has been suggested that polyamines may be considered precocious markers of rooting. Finally, the root growth-related effects of polyamines on some species, such as olive, are due to the low endogenous polyamine content compared to other species, such as walnut and chestnut, which are not affected by polyamine treatment.

EFFECT OF PLANT GROWTH REGULATORS IN ROOT INITIATION

Adventitious Roots

Adventitious rooting may be defined as the formation of roots at different locations than where roots occur under natural conditions. This phenomenon is particularly useful for the asexual propagation of plant material from cuttings. Plant parts such as stems, roots, or leaves can serve as a source for adventitious root formation (Arteca, 1996).

The origin of adventitious roots varies in relation to the plant part used (Arteca, 1996). In stem cuttings, following periderm formation, cells near the vascular cambium and phloem will divide and subsequently form adventitious roots. In leaf cuttings, roots can develop from primary and secondary meristems. Numerous factors contribute to this process; plant growth regulators are one of the most important.

It was postulated that a substance, termed rhizocaline, found in leaves, buds, and/or cotyledons, was able to move to the roots and stimulate rooting (Bouillene and Went, 1933). Today, rhizocaline still remains a hypothetical compound. However, different classes of plant growth regulators have been proven to influence root initiation, including auxins, cytokinins, gibberellins, ethylene, and brassinolide, as well as inhibitory substances such as abscisic acid, growth retardants, and phenolics. To date, auxins have been shown to have the greatest effect on rooting. Numerous reports have indicated the involvement of auxin in the initiation of adventitious roots, and that division of root initials is dependent upon exogenous or endogenous auxin. Synthetic auxins, such as indole-3-butyric acid (IBA) and naphthaleneacetic acid (NAA), have been shown to be more effective than the naturally occurring IAA for rooting (Thimann, 1935a). In addition, many studies have hypothesized a role for polyamines in the rooting process, and their relationship with auxins and peroxidases (Friedman, Altman, and Bachrach, 1982; 1985; Altamura, 1994). It has been suggested that putrescine controls the IAA level, mediated by some peroxidases, but the reverse is also true (Gaspar, Kevers, and Hausman, 1997). According to Gaspar and colleagues (1997), it might be hypothesized that IAA and putrescine, which are known to control cell division cycles (Del Duca and Serafini-Fracassini, 1993), are required to initiate cell divisions at the end of the rooting inductive phase.

Also, ethylene and ethylene analogs have been shown to stimulate adventitious root formation (Zimmerman and Hitchcock, 1933). This may be correlated with the fact that auxins stimulate ethylene biosynthesis (Arteca, 1990), and it has been suggested that auxin- induced ethylene, rather than auxin itself, may induce adventitious root formation (Mudge, 1989). On the other hand, cytokinins and gibberellins have been shown to inhibit root formation in many species (Arteca, 1996). In regard to ABA and brassinosteroids, their significance in adventitious root formation remains unknown.

Lateral Roots

Lateral roots and primary roots are almost identical in organization. Therefore, it is reasonable to predict that both organs exhibit similar developmental processes (Malamy and Benfey, 1997). Most species have two cell types in the root epidermis, cells that produce long, hair-like extensions (root hair cells, which are derived from trichoblasts) and hairless cells (derived from atrichoblasts) (Cutter, 1978). Some plants show no apparent pattern of root hair and hairless cells (Leavit, 1904). In other plants, including many monocots, epidermal cell fate is linked to an asymmetric cell division, in which the smaller daughter cell differentiates into a root hair cell and the larger daughter cell generates one or more mature hairless cells (Cutter and Feldman, 1970). In a third group that includes members of the Brassicaceae, such as Arabidopsis, a distinct position-dependent pattern of epidermal cell types is generated (Galway et al., 1994). In this system, trichoblasts form in the crevice between underlying cortical cells, whereas atrichoblasts develop over a cortical cell.

Lateral roots in Arabidopsis are initiated in pericycle cells immediately contiguous to the two protoxylem poles (Laskowski et al., 1995). Although all pericycle cells are identical, not all of them are able to form lateral roots (Malamy and Benfey, 1997). Several mechanisms have been postulated to explain this, such as a developmental cell fate. However, exogenous application of auxin, the removal of the root tip, or the presence of certain ions, such as zinc (Zn), tend to greatly increase the number of lateral roots (Torrey, 1986; Davies, Francis, and Thomas, 1991). Contrarily, the lack of metal ions decreases the size of root meristems, a phenomenon regarded as "chemical decapitation" (Cadiz, Davies, and de Guzman, 1995).

Studies with Arabidopsis mutants have been useful in correlating root hair formation with the action of plant hormones, specifically auxin and ethylene. Evidence of ethylene activity relies on mutations affecting the *constitutive triple response1* (CTR1) locus, which encodes a Raf-like protein kinase (proposed to negatively regulate the ethylene signal transduction pathway) and causes root hairs to form on epidermal cells that normally are hairless (Dolan et al., 1994). The hairless phenotype of the *rhd6* (root hair defective 6) mutant can be suppressed by the supply of ACC (ethylene precursor) or IAA to the media (Masucci and Schiefelbein, 1994). A GCA box (*cis*-acting auxin-response element) has been identified in the *LRP1* (lateral root primordium) gene (Smith and Federoff, 1995). The presence of this GCA box suggests that the expression of *LRP1* may be regulated by auxin during early lateral root primordium development.

Enough information exists to suggest that auxin is strongly involved in lateral root formation. This phenomenon has been corroborated in the *sur* (superroot) and *rty* (rooty) Arabidopsis mutants, which have increased concentrations of IAA compared with wild- type plants (Boerjan et al., 1995; King et al., 1995). Transgenic modified plants, which overexpress auxin, have been shown to exhibit the same phenotype (Klee, 1987). Contrarily, the auxin-resistant mutants *axr1*, *axr4*, and *aux1* show a dramatic reduction in lateral root formation (Hobbie and Estelle, 1995; Timpte et al., 1995). Mutants that inhibit the transport of auxin from the upper parts of the plant, such as *tir* (transport inhibitor resistant), show a reduced number of lateral roots as well (Ruegger et al., 1997). The existence of lateral root-specific genes has been indicated by the *alf4* (aberrant lateral root formation) mutant in Arabidopsis (Celenza, Grisafi, and Fink, 1995). The *alf4* mutant completely lacks the ability to produce lateral roots and cannot be rescued by the addition of auxin or auxin precursors. However, the mechanism by which auxin triggers lateral root initiation and formation in specific pericycle cells is still unknown.

The isolation of Arabidopsis mutants with altered root gravitropism has implicated auxin as the plant hormone involved in the gravitropic response (Benfey and Schiefelbein, 1994). Normally, Arabidopsis plants, and plants in general, direct their growth toward a gravity vector. However, mutations in any of at least five loci (*aux1, dwf, agr1, axr1,* and *axr2*) cause plant roots to respond abnormally to gravity or not to respond at all (Mirza et al., 1984; Estelle and Somerville, 1987; Bell and Maher, 1990; Wilson et al., 1990). Four of these mutants (*dwf, aux1, axr1,* and *axr2*) were shown to exhibit an auxin-resistant phenotype.

Future studies will provide insights into the mechanisms by which auxin and ethylene promote lateral root initiation and formation, and the molecular and biochemical implications of these processes.

Agrobacterium Rhizogenes "Hairy Root" Formation

The soil bacterium *Agrobacterium rhizogenes* causes "hairy root" disease in dicotyledonous plants, a disease characterized by the proliferation of adventitious roots at the infection site. The infection process is characterized by the integration of a portion of the Ri plasmid, called T-DNA, into the plant genome (Chilton et al., 1982). This event is mediated by the virulence (vir) genes within the Ri plasmid.

The expression of the vir genes is mediated by the phenolic compounds, such as α-hydroxyacetosyringone (HOAS) and acetosyringone. Inositol has been shown to enhance the activity of HOAS, while glucose, rhamnose, and xylose stimulate the effect of acetosyringone (Yang- Nong et al., 1990).

The T-DNA genes *rolA, rolB,* and *rolC* have been shown to be involved in "hairy root" formation (Spena et al., 1987). The exact role of these genes is still unclear and may vary among plant species. In tobacco, *rolA* is mainly responsible for the development of "hairy roots," while *rolB* appears to be a factor in "hairy root" initiation (Cardarelli et al., 1987). Transgenic tobacco plants expressing the *rolB* gene have increased auxin activity (Estruch, Schell, and Spena, 1991; Nilsson, Crozier, et al., 1993). The rolB protein has been proposed to be a ß-glucosidase, acting by hydrolyzing bound auxins (Estruch, Schell, and Spena, 1991). The action of the *rolB* gene may thus lead to an increase in the intracellular concentration of free IAA. Expression of *rolC* in transgenic tobacco plants resulted in a reduced concentration of isopentenyladenosine (iPA) and increased levels of GA_{19} (Estruch et al., 1991; Nilsson, Moritz, et al., 1993). Transgenic potato and tobacco plants expressing *rolA* or derivatives of *rolA* and *rolC* had reduced gibberellic acid levels (Dehio et al., 1993).

Transformed hairy root cultures provide a fascinating system to study the biology, biochemistry, and molecular biology of root-related processes, such as root growth regulation (Flores and Curtis, 1992). Moreover, the hairy root phenotype is stable in culture (see Figure 1.3), and the growth rate is much faster than that of normal roots (Rhodes et al., 1990).

FUTURE PERSPECTIVES

It should be clear from the previous discussion that some basic knowledge exists concerning the role of plant growth regulators in root regulatory processes. However, many gaps still should be filled in, such as the hormonal mechanism associated with storage root formation. Tuber formation in vitro is now a routine procedure for many potato genotypes (Pelacho and Mingo-Castel, 1991). However, none of the growth regulators (cytokinins, jasmonic acid) that facilitate in vitro tuberization has so far proven effective for induction of storage roots in culture (Flores, unpublished data). The discovery of such a signal will greatly enhance our knowledge of root formation and our ability to understand and manipulate root and overall plant physiology.

The use of *Agrobacterium rhizogenes*-transformed "hairy roots" has been proven to be effective in the study and manipulation of root secondary metabolites (Flores and Curtis, 1992). This system could also be utilized in the study of root growth regulatory mechanisms and in the discovery of novel plant hormones. Finally, the availability of root developmental mutants isolated from Arabidopsis is beginning to provide some insights on root growth and development processes (Smith and Federoff, 1995).

FIGURE 1.3. Hairy Root Cultures of *Mirabilis longiflora*

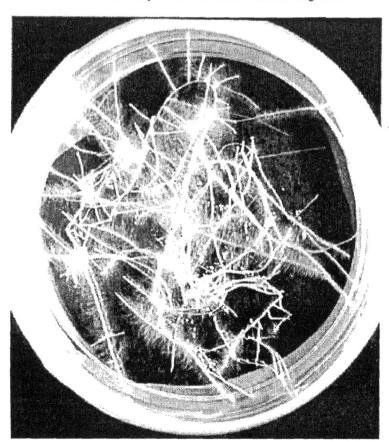

REFERENCES

Abeles, F.B., P.W. Morgan, and M.E. Saltveit Jr. (1992). *Ethylene in Plant Biology,* Second Edition. San Diego, CA: Academic Press.

Abou-Mandour, A.A. and W. Hartung (1980). The effect of abscisic acid on growth and development of intact seedlings, root and callus cultures and stem and root segments of *Phaseolus coccineus. Zeitschrift für Pflanzenphysiologie* 100:25.

Aldington, S. and S. Fry (1993). Oligosaccharins. *Advances in Botanical Research* 19:1101.

Altamura, M.M. (1994). Rhizogenesis and polyamines in tobacco thin cell layers. *Advances in Horticultural Sciences* 8:33.

Arteca, R. (1990). Hormonal stimulation of ethylene biosynthesis. In *Polyamines and Ethylene: Biochemistry, Physiology, and Interactions,* H.E. Flores, R.N. Arteca, and J.C. Shanon (Eds.). Rockville, MD: American Society of Plant Physiologists, pp. 216-223.

Arteca, R. (1996). Plant Growth Substances: Principles and Applications. New York: Chapman and Hall.

Bagni, N., D. Serafini-Frascassini, and P. Torrigiani (1982). Polyamines and cellular growth processes in higher plants. In *Plant Growth Substances,* P.F. Wareing (Ed.). London: Academic Press, pp. 473-482.

Bell, C.J. and E.P. Maher (1990). Mutants of *Arabidopsis thaliana* with abnormal gravitropic responses. *Molecular and General Genetics* 220:289-293.

Benfey, P.N. and J. Schiefelbein (1994). Getting to the root of plant development: The genetics of Arabidopsis root formation. *Trends in Genetics* 10:84-88.

Ben-Gad, D.Y., A. Altman, and S.P. Monselise (1979). Interrelationships of vegetative growth and assimilate distribution of *Citrus limettioides* seedlings in response to root-applied GA$_3$ and SADH. *Canadian Journal of Botany* 57:484-490.

Berger, S., E. Bell, and J.E. Mullet (1996). Two methyl jasmonate-insensitive mutants show altered expression of Atvsp in response to methyl jasmonate and wounding. *Plant Physiology* 111:525-531.

Bialek, K., L. Michalczuk, and J.D. Cohen (1992). Auxin biosynthesis during seed germination in *Phaseolus vulgaris. Plant Physiology* 100:509-517.

Biondi, S., T. Diaz, I. Iglesias, G. Gamberini, and N. Bagni (1990). Polyamine and ethylene in relation to adventitious root formation in *Prunus avium* shoot cultures. *Physiologia Plantarum* 78:474-483.

Boerjan, W., M.T. Cervera, M. Delarue, T. Beeckman, W. Dewitte, C. Bellini, M. Caboche, M. Van Onckelen, M. Van Montagu, and D. Inze (1995). *Superroot,* a recessive mutation in Arabidopsis, confers auxin overproduction. *The Plant Cell* 7:1405-1419.

Bouillenne, R. and F.W. Went (1933). Recherches experimentales sur la neoformation des racines dans les plantules et les boutures des plantes superieures. *Annuel Jardin Botanie Buitenzorg* 43:25-202.

Bradford, K.J. and S.F. Yang (1980). Xylem transport of 1-aminocyclo-propane-1-carboxylic acid, an ethylene precursor in waterlogged tomato plants. *Plant Physiology* 65:322-336.

Brown, P.H. and T.-H.D. Ho (1986). Barley aleurone layers secrete a nuclease in response to gibberellic acid. Purification and partial characterization of the associated ribonuclease, deoxyribonuclease, and 3' nucleotide activities. *Plant Physiology* 82:801-806.

Cadiz, N.M., M.S. Davies, and C.C. de Guzman (1995). Root growth characteristics of *Ocimum* L. (holy basil) and *Festuca rubra* L. cv. Merlin (red fescue) in response to cadmium, lead and zinc. *The Philippine Agriculturist* 78:331-342.

Cardarelli, M., D. Marriotti, M. Pomponi, L. Spano, I. Capone, and P. Constantino (1987). *Agrobacterium rhizogenes* T-DNA genes capable of inducing hairy root phenotype. *Molecular and General Genetics* 209:475-480.

Celenza, J.L., P.L. Grisafi, and G.R. Fink (1995). A pathway for lateral root formation in *Arabidopsis thaliana. Genes & Development* 9:2131-2142.

Chen, C.M., J.R. Ertl, S.M. Leisner, and C.C. Chang (1985). Localization of cytokinin biosynthetic sites in pea plants and carrot roots. *Plant Physiology* 78:510-513.

Chilton, M.D., D.A. Tepfer, A. Petit, C. David, F. Casse-Delbart, and J. Tempe (1982). *Agrobacterium rhizogenes* inserts T-DNA into the genomes of the host plant root cells. *Nature* 295:432-434.

Cohen, J.D. and K. Bialek (1984). The biosynthesis of indole-3-acetic acid in higher plants. In *The Biosynthesis and Metabolism of Plant Hormones*, A. Crozier and J.R. Hillman (Eds.). Cambridge: Cambridge University Press, pp. 165-181.

Cohen, S.S. (1971). *Introduction to the Polyamines.* Englewood Cliffs, NJ: Prentice-Hall.

Cornish, K. and J.A.D. Zeevaart (1985). Abscisic acid accumulation by roots of *Xanthium strumarium* L. and *Lycopersicon esculentum* Mill. in relation to water stress. *Plant Physiology* 79:653-658.

Crozier, A. and D.M. Reid (1971). Do roots synthesize gibberellins? *Canadian Journal of Botany* 49:967-975.

Cutter, E.G. (1978). The epidermis. In *Plant Anatomy.* London: Clowes and Sons, pp. 94-106.

Cutter, E.G. and L.J. Feldman (1970). Trichoblasts in Hydrocharis. I. Origin, differentiation, dimensions and growth. *American Journal of Botany* 57:190-201.

Davies, M.S., D. Francis, and J.D. Thomas (1991). Rapidity of cellular changes induced by zinc in a zinc tolerant and non-tolerant cultivar of *Festuca rubra* L. cv. Merlin. *New Phytology* 117:103-108.

Davies, W.J. and H.G. Jones (1991). *Abscisic Acid: Physiology and Biochemistry.* Oxford: Bios Scientific Publishers.

Dehio, C., K. Grossman, J. Schell, and T. Schmulling (1993). Phenotype and hormonal status of transgenic tobacco plants overexpressing the rolA gene of *Agrobacterium rhizogenes* T-DNA. *Plant Molecular Biology* 23:1199-1210.

Del Duca, S. and D. Serafini-Fracassini (1993). Polyamines and protein modification during the cell cycle. In *Molecular and Cell Biology of the Plant Cell Cycle*, J.C. Ormrod and D. Francis (Eds.). Dordrecht, Netherlands: Kluwer Academic Publishers.

Desai, H.V. and A.R. Mehta (1984). Changes in polyamine levels during shoot formation, root formation and callus induction in cultured *Passiflora* leaf discs. *Plant Physiology* 119:45-53.

Dolan, L., C. Duckett, C. Grierson, P. Linstead, K. Schneider, E. Lawson, C. Dean, R.S. Poethig, and K. Roberts (1994). Clonal relations and patterning in the root epidermis of *Arabidopsis. Development* 120:2465-2474.

Dong, C.N. and R.N. Arteca (1982). Changes in photosynthetic rates and growth following root treatments of tomato plants with phytohormones. *Photosynthesis Research* 3:45-52.

Elliott, M.C. (1977). Auxins and the regulation of root growth. In *Proceedings 9th International Conference, Plant Growth Substances,* Lausanna, 1976, P.E. Pilat (Ed.). Berlin: Springer-Verlag.

Eshel, A. and Y. Waisel (1997). Aeroponics: A search for understanding roots. In *Biology of Root Formation and Development,* A. Altman and Y. Waisel (Eds.). New York: Plenum Press, New York.

Estelle, M.A. and C. Somerville (1987). Auxin-resistant mutants of *Arabidopsis thaliana* with an altered morphology. *Molecular and General Genetics* 206:200-206.

Estruch, J.J., D. Chiriqui, K. Grossman, J. Schell, and A. Spena (1991). The plant oncogene rolC is responsible for the release of cytokinins from glucoside conjugates. *European Molecular Biology Organization Journal* 10:2889-2895.

Estruch, J.J., J. Schell, and A. Spena (1991). The protein encoded by the rolB plant oncogene hydrolyses indole glucosides. *European Molecular Biology Organization Journal* 10:3125-3128.

Feldman, L.J. (1980). Auxin biosynthesis and metabolism in isolated roots of *Zea mays. Physiologia Plantarum* 49:145-150.

Feldman, L.J. (1982). Formation and partial characterization of growth inhibitors from cultured and intact root caps. *Annals of Botany* 50:747-756.

Flores, H.E. and W.R. Curtis (1992). Approaches to understanding and manipulating the biosynthetic potential of plant roots. *Annals New York Academy of Science* 665:188-209.

Flores, H.E., C.M. Protacio, and M.W. Signs (1989). Primary and secondary metabolism of polyamines in plants. *Recent Advances in Phytochemistry* 23:329-393.

Flores, H.E., N.D. Young, and A.W. Galston (1984). Polyamine metabolism and plant stress. In *Cellular and Molecular Biology of Plant Stress,* J.L. Key and T. Kosuge (Eds.). New York: Allan R. Liss.

Friedman, R., A. Altman, and U. Bachrach (1982). Polyamines and root formation in mung bean hypocotyl cuttings. I. Effects of exogenous compounds and changes in endogenous polyamine content. *Plant Physiology* 70:844.

Friedman, R., A. Altman, and U. Bachrach (1985). Polyamines and root formation in mung bean hypocotyl cuttings. II. Incorporation of precursors into polyamines. *Plant Physiology* 79:80.

Fries, N. (1960). The effect of adenine and kinetin on growth and differentiation of *Lupinus. Physiologia Plantarum* 13:468.

Galway, M.E., J.D. Masucci, A.M. Lloyd, V. Walbot, R.W. Davis, and J.W. Schiefelbein (1994). The TTG gene is required to specify epidermal cell fate and cell patterning in the *Arabidopsis* root. *Developmental Biology* 166:740-754.

Gaspar, T., C. Kevers, and J.-F. Hausman (1997). Indissociable chief factors in the inductive phase of adventitious rooting. In *Biology of Root Formation,* A. Altman and Y. Waisel (Eds.). New York: Plenum Press.

Gaudin, V., T. Vrain, and L. Jouanin (1994). Bacterial genes modifying hormonal balances in plants. *Plant Physiology and Biochemistry* 32:11-29.

Gross, D. and B. Parthier (1994). Novel natural substances acting in plant growth regulation. *Journal of Plant Growth Regulation* 13:93-114.

Grove, M., G. Spencer, W. Rohwedder, N. Mandava, J. Worley, J. Warthan, G. Stephens, J. Flippen-Anderson, and J. Cook (1979). Brassinolide, a plant growth-promoting steroid isolated from *Brassica napus* pollen. *Nature* 281: 216-217.

Haagen-Smit, A.J., W.B. Dandliker, S.H. Wittwer, and A.E. Murneek (1946). Isolation of 3-indolacetic acid from immature corn kernels. *American Journal of Botany* 33:118-120.

Hamilton, A.J., G.W. Lycett, and D. Grierson (1990). Antisense gene that inhibits synthesis of the hormone ethylene in transgenic plants. *Nature* 346:284-287.

Hartung, W. and A.A. Abou-Mandour (1980). Abscisic acid in root culture of *Phaseolus coccineus* L. *Zeitschrift für Pflanzenphysiologie* 97:265-269.

Hartung, W. and N.C. Turner (1997). Abscisic acid relations in stressed roots. In *Biology of Root Formation and Development*, A. Altman and Y. Waisel (Eds.). New York: Plenum Press.

Hisamatsu, T., M. Koshioka, S. Kubota, T. Nishijima, H. Yamane, R.W. King, and L.N. Mander (1998). Isolation and identification of GA_{112} (12β-hydroxy-GA_{12}) in *Matthiola incana*. *Phytochemistry* 47:3-6.

Hobbie, L. and M. Estelle (1995). The *axr4* auxin-resistant mutants of *Arabidopsis thaliana* define a gene important for root gravitropism and lateral root initiation. *Plant Journal* 7:211-220.

Ingram, T.J., J.B. Reid, I.C. Murfet, P. Gaskin, C.L. Willis, and J. MacMillan (1984). Internode length in *Pisum*. The Le gene controls the 3β-hydroxylation of gibberellin A_{20} to gibberellin A_1. *Planta* 160:455-463.

Itai, C. and H. Birnbaum (1996). Synthesis of plant growth regulators by roots. In *Plant Roots: The Hidden Half*, Y. Waisel, A. Eshel, and U. Kafkafi (Eds.). New York: Marcel Dekker.

Kaminek, M., D.W.S. Mok, and E. Zazimalova (1992). *Physiology and Biochemistry of Cytokinins in Plants*. Hague: SPB Academic Publishing.

Kende, H. (1993). Ethylene biosynthesis. *Annual Review of Plant Physiology and Plant Molecular Biology* 44:283-307.

Kende, H. and J.A.D. Zeevaart (1997). The five "classical" plant hormones. *The Plant Cell* 9:1197-1210.

King, J.J., D.P. Stimart, R.H. Fisher, and A.B. Bleecker (1995). A mutation altering auxin homeostasis and plant morphology in Arabidopsis. *The Plant Cell* 7:2023-2037.

Kionka, C. and N. Amrhein (1984). The enzymatic malonylation of 1-aminocyclopropane-1-carboxylic acid in homogenates of mung-bean hypocotyls. *Planta* 162:226-235.

Klee, H.J. and C.P. Romano (1994). The roles of phytohormones in development as studied in transgenic plants. *Critical Review in Plant Sciences* 13:311-324.

Klee, J.H. (1987). The effects of overproduction of two *Agrobacterium tumefaciens* T-DNA auxin biosynthetic gene products in transgenic petunia plants. *Genes & Development* 1:86-96.

Koda, Y. (1992). The role of jasmonic acid and related compounds in the regulation of plant development. *International Review of Cytology* 135:155-199.

Kogl, F. and A.J. Haagen-Smit (1931). Uber die Chemie des Wuchsstoffs K. Akad. Wetenschap. Amsterdam. *Proc. Sect. Sci.* 34:1411-1416.

Kogl, F. and D.G.F.R. Kostermans (1934). Heteroauxin als Stoffwechselprodukt niederer pflanzlicher Organism, Isolierung aus Hefe. *Zeitschrift Physiologie Chemie* 225:215-229.

Laskowski, M.J., M.E. Williams, H.C. Nusbaum, and I.M. Sussex (1995). Formation of lateral root meristems is a two stage process. *Development* 121:3303-3310.

Leavit, R.G. (1904). Trichomes of the root in vascular cryptograms and angiosperms. *Proceedings Boston Society of Natural History* 31:273-313.

Letham, D.S. and L.M.S. Palni (1983). The biosynthesis and metabolism of cytokinins. *Annual Review of Plant Physiology* 34:163-197.

Malamy, J.E. and P.N. Benfey (1997). Down and out in Arabidopsis: The formation of lateral roots. *Trends in Plant Science* 2:390-396.

Martin, R.C., M.C. Mok, and D.W.S. Mok (1993). Cytolocalization of zeatin O-xylosyltransferase in *Phaseolus*. *Proceedings of the National Academy of Sciences USA* 90:953-957.

Masucci, J.D. and J.W. Schiefelbein (1994). The *rhd6* mutation of *Arabidopsis thaliana* alters root-hair initiation through an auxin and ethylene associated process. *Plant Physiology* 106:1335-1346.

Metzger, J.D. and J.A.D. Zeevaart (1980). Comparison of the levels of six endogenous gibberellins in roots and shoots of spinach in relation to photoperiod. *Plant Physiology* 66:679-683.

Mirkovic, K. (1993). Effect of polyamines on rooting of hardy deciduous (*Rhododendron* sp.) in vitro. Master's thesis, Mediterranean Agronomic Institute of Chania, Crete, Greece.

Mirza, J.I., G.M. Olsen, T.-H. Iversen, and E.P. Maher (1984). The growth and gravitropic responses of wild type and auxin-resistant mutants of *Arabidopsis thaliana*. *Physiologia Plantarum* 60:516-522.

Motyka, V., M. Faiss, M. Strnad, M. Kaminek, and T. Schmulling (1996). Changes in cytokinin content and cytokinin oxidase activity in response to derepression of ipt gene transcription in transgenic tobacco calli and plants. *Plant Physiology* 112:1035-1043.

Mudge, K.W. (1989). Effect of ethylene on rooting. In *Adventitious Root Formation in Cuttings*, T.D. Davies, B.E. Haissig, and N. Sankhla (Eds.). Portland, OR: Dioscorides Press.

Muir, R.M. and B.P. Lantican (1968). Purification and properties of the enzyme system forming indolacetic acid. In *Biochemistry and Physiology of Plant Growth Substances*, F. Wightman and G. Setterfield (Eds.). Ottawa: Runge Press, pp. 259-272.

Nilsson, O., A. Crozier, T. Schmulling, G. Sandberg, and O. Olsson (1993). Indole-3-acetic acid homeostasis in transgenic tobacco plants expressing the rolB gene. *Plant Journal* 3:681-689.

Nilsson, O., T. Moritz, N. Imbault, G. Sandemberg, and O. Olsson (1993). Hormonal characterization of transgenic tobacco plants expressing the rolC gene of *Agrobacterium rhizogenes* T_L-DNA. *Plant Physiology* 102:363-371.

Normanly, J., J.P. Slovin, and J.D. Cohen (1995). Rethinking auxin biosynthesis and metabolism. *Plant Physiology* 107:323-329.

Olson, D.C., J.A. White, L. Edelman, R.N. Harkins, and H. Kende (1991). Differential expression of two genes for 1-aminocyclopropane-1-carboxylate synthase in tomato fruits. *Proceedings of the National Academy of Sciences USA* 88:5340-5344.

Parthier, B., C. Bruckner, W. Dathe, B. Hause, G. Herrman, H.-D. Knofel, R. Kramell, J. Lehmann, O. Miersch, S.T. Reinbothe, et al. (1992). Jasmonates: Metabolism, biological activities, and modes of action in senescence and stress responses. In *Progress in Plant Growth Regulation,* C.M. Karssen, L.C. van Loon, and D. Vreugdenhil (Eds.). Dordrecht, Netherlands: Kluwer Academic Publishers, pp. 276-285.

Pelacho, A.M. and A.M. Mingo-Castel (1991). Jasmonic acid induces tuberization of potato stolons cultured in vitro. *Plant Physiology* 97:1253-1255.

Raskin, I. (1992). Role of salicylic acid in plants. *Annual Review of Plant Physiology and Plant Molecular Biology* 43:439-463.

Ravnikar, M., B. Vilhar, and N. Gogala (1992). Stimulatory effects of jasmonic acid on potato stem node and protoplast culture. *Journal of Plant Growth Regulation* 11:29-33.

Rhodes, M.J., R.J. Robins, J.D. Hamill, A.J. Parr, M.J. Hilton, and N.J. Walton (1990). Properties of transformed root cultures. In *Secondary Products from Plant Tissue Culture,* B.V. Charlwood and M.J. Rhodes (Eds.). Oxford: Clarendon Press, pp. 201-225.

Rivier, L., H. Milon, and P.E. Pilet (1977). Gas chromatography-mass spectromeric determinations of abscisic acid levels in the cap and the apex of maize roots. *Planta* 134:23.

Roddick, J.G. and M. Guan (1991). Brassinosteroids and root development. In *Brassinosteroids. Chemistry, Bioactivity, and Applications,* H.G. Cutler, T. Yokota, and G. Adam (Eds.). Washington, DC: American Chemical Society, pp. 231-245.

Ross, J.J., J.B. Reid, S.M. Swain, O. Hasa, A.T. Poole, P. Hedden, and C.L. Willis (1995). Genetic regulation of gibberellin deactivation in *Pisum. Plant Journal* 7:513-523.

Ruegger, M., E. Dewey, L. Hobbie, D. Brown, P. Bernasconi, J. Turner, G. Mudday, and M. Estelle (1997). Reduced NPA-binding in the *tir3* mutant of Arabidopsis is associated with a reduction in polar auxin transport and diverse morphological defects. *The Plant Cell* 9:745-757.

Rugini, E., G. Di Francesco, M. Muganu, S. Astolfi, and G. Caricato (1997). The effect of polyamines and hydrogen peroxide on root formation in olive and the role of polyamines as early marker for rooting ability. In *Biology of Root Formation and Development,* A. Altman and Y. Waisel (Eds.). New York: Plenum Press.

Rugini, E. and X.S. Wang (1986). Effect of polyamines, 5-azacytidine and growth regulators on rooting in vitro of fruit trees, treated and untreated with *Agrobacterium rhizogenes. Proceedings International Congress of Plant Tissue and Cell Culture.* Minnesota, p. 374.

Saab, I.N., R.E. Sharp, J. Pritchard, and G.S. Voetburg (1990). Increased endogenous abscisic acid maintains primary root growth and inhibits shoot growth of maize seedlings at low water potential. *Plant Physiology* 93:1329.

Sakurai, A. and S. Fujioka (1994). The current status of physiology and biochemistry of brassinosteroids: A review. *Journal of Plant Growth Regulation* 13:147-159.

Salkowski, E. (1885). Uber das verhalten der skatolcarbonsaure im organismus. *Zeitschrift für Physiologie Chemie* 9:23-33.

Sanders, I.O., A.R. Smith, and M.A. Hall (1989). Ethylene metabolism in *Pisum sativum* L. *Planta* 179:104-114.

Sembdner, G. and B. Parthier (1993). The biochemistry and the physiological and molecular actions of jasmonates. *Annual Review of Plant Physiology and Plant Molecular Biology* 44:569-589.

Slocum, R.D., R. Kaur-Sawhney, and A.W. Galston (1984). The physiology and biochemistry of polyamine in plants. *Archives of Biochemistry and Biophysics* 235(2):283-303.

Smith, D.L. and N.V. Federoff (1995). LRP1, a gene expressed in lateral and adventitious root primordia in Arabidopsis. *The Plant Cell* 7:735-745.

Spena, A., T. Schmulling, C. Kones, and J. Schell (1987). Independent and synergistic activity of rolA, B and C loci in stimulating abnormal growth in plants. *European Molecular Biology Organization Journal* 6:3891-3899.

Staswick, P.E. (1992). Jasmonate, genes, and fragrant signals. *Plant Physiology* 99:804-807.

Stendil, G. (1982). Cytokinins as inhibitors of root growth. *Physiologia Plantarum* 56:500-506.

Takahashi, N., B.O. Phinney, and J. MacMillan (1991). *Gibberellins*. Berlin: Springer-Verlag.

Thimann, K.V. (1935a). On an analysis of activity of two growth-promoting substances of plant tissues. *Proc. Kon. Ned. Akad. Wet.* 38:896-912.

Thimann, K.V. (1935b). On the plant growth hormone produced by *Rhizopus suimus. Journal of Biological Chemistry* 109:279-291.

Tietz, A. (1971). Nachweis von abscisinsaure in wurzein. *Planta* 96:93.

Timpte, C., C. Lincoln, F.B. Pickett, J. Turner, and M. Estelle (1995). The *axr1* and *aux1* genes of Arabidopsis function in separate auxin-response pathways. *Plant Journal* 8:561-569.

Torrey, J.G. (1976). Root hormones and plant growth. *Annual Review of Plant Physiology* 24:435-459.

Torrey, J.G. (1986). Endogenous and exogenous influences on the regulation of lateral root formation. In *New Root Formation in Plants and Cuttings,* M.B. Jackson and M. Nijhoff (Eds.). Boston, MA: Kluwer Academic Publishers, pp. 31-66.

van den Berg, J.H. and E.E. Ewing (1991). Jasmonates and their role in plant growth and development, with special reference to the control of potato tuberization: A review. *American Potato Journal* 68:781-794.

Wample, R.L. (1979). The role of endogenous auxins and ethylene in the formation of adventitious roots and hypocotyl hypertrophy in flooded sunflower plants (*Helianthus annus*). *Physiologia Plantarum* 45(2):227-234.

Went, F.W. and K.V. Thimann (1937). Phytohormones. New York: Macmillan.

Wildman, S.G., M.G. Ferri, and J. Bonner (1947). The enzymatic conversion of tryptophan to auxin by spinach leaves. *Archives of Biochemistry and Biophysics* 13:131.

Wilson, A.K., F.B. Pickett, J.C. Turner, and M. Estelle (1990). A dominant mutation in Arabidopsis confers resistance to auxin, ethylene, and abscisic acid. *Molecular and General Genetics* 222:377-383.

Yabuta, T. (1935). Biochemistry of the bakanae fungus on rice. *Agronomy & Horticulture* (Tokyo) 10:17-22.

Yang, S.F. and Hoffman, N.E. (1984). Ethylene biosynthesis and its regulation in higher plants. *Annual Review of Plant Physiology* 35:155-189.

Yang-Nong, S., M. Shibuya, Y. Ebizuka, and U. Sankawa (1990). Hydroxyacetosyringone is the major virulence gene activating factor in belladonna hairy root cultures, and inositol enhances its activity. *Chemical and Pharmaceutical Bulletin* 38:2063-2065.

Yip, W.K., T. Moore, and S.F. Yang (1992). Differential accumulation of transcripts for four tomato 1-aminocyclopropane-1-carboxylate synthase homologs under various conditions. *Proceedings of the National Academy of Sciences USA* 89:2475-2479.

Yokota, T. (1997). The structure, biosynthesis and function of brassinosteroids. *Trends in Plant Science* 2:137-143.

Zavala, M.E., and D.L. Brandon (1983). Localization of a phytocrome using immunocytochemistry. *Journal of Cell Biology* 97:1235-1239.

Zimmerman, P.W. and A.E. Hitchcock (1933). Initiation and stimulation of adventitious roots caused by unsaturated hydrocarbon gases. *Contribution of Boyce Thompson Institute* 5:351-369.

Chapter 2

Manipulating Yield Potential in Cereals Using Plant Growth Regulators

Ari Rajala
Pirjo Peltonen-Sainio

PLANT GROWTH REGULATORS IN CEREAL PRODUCTION: HISTORY AND PRESENT DAY IN A NUTSHELL

Plant growth regulators (PGRs) are traditionally and almost exclusively used in high-input cereal management to shorten the straw and, consequently, increase lodging resistance. Lodging often results in serious problems in grain growth. It may cause disharmony in source-sink interaction at different levels: (1) in the ability of the green area to capture photosynthetically active radiation (PAR), (2) in the radiation-use efficiency (RUE), which expresses the biomass accumulated per unit of intercepted radiation, and (3) in the partitioning of dry matter between harvestable organs and nonyield structures. The lodged plant stand is evidently far from the ideal crop canopy, and, furthermore, lodging interferes with translocation of, for example, water, nutrients, de novo synthesized photosynthate, and carbohydrates mobilized from the stem and additional reserves. How drastically grain-filling rate is reduced depends on the extent of the lodging and the ability of the plant stand to rise again and recover. Additionally, lodging occurring close to harvest leads to troubles in combine har-

The authors wish to thank the Finnish Academy of Sciences for financing their work with PGRs.

27

vesting and, in conjunction with high precipitation, often causes deterioration of the grain quality. This is typical for northern European growing conditions that are characterized by high precipitation close to harvest, and especially critical in modern, relatively sprouting-susceptible bread wheat *(Triticum aestivum* L.) and malting barley *(Hordeum vulgare* L.) cultivars that lose their dormancy already in the intact heads. Thus, PGRs improve grain quantity and quality indirectly rather than directly and enhance photosynthate production and partitioning into grain.

The history of PGRs in manipulating growth of small-grain cereals (later referred to only as cereals)—more precisely stem elongation—is brief, covering exactly four decades. Agriculture is some 10,000 years old. If we condense this period into twenty-four hours, we have used science-based plant breeding for some fifteen minutes, and PGRs in cereals for about six minutes. If we change the timescale such that humans have been on the earth for twenty-four hours, we have practiced agriculture for some five minutes, used science-based breeding for three seconds, and PGRs in cereals for about one second. Domestication of cereals—wheat and barley in the early stages—was based on improvements in characteristics associated with enhanced yielding ability when grown in dense stands. From the very beginning of science-based plant breeding, and evidently even long before that, shortening and stiffening the straw, thereby enabling the straw to support the continuously increasing head weight, was one of the most appreciated modifications of a nonyield component. As agriculture became more and more mechanized and scientists demonstrated increasing possibilities for using inputs to enhance grain production, the plant breeders' achievements in shortening the straw and preventing lodging were partly lost. This is documented, for example, in oat *(Avena sativa* L.) grown in northern growing conditions, by Peltonen-Sainio and colleagues (1993). Thus, additional methods were needed to exploit fully the crop's yield potential, and PGRs enabled the use of high seeding rates and high nitrogen fertilizer application rates, in particular.

The first PGR used to control stem elongation in cereals was chlormequat chloride (CCC). It was tested and evaluated in the late 1950s and the 1960s by Tolbert (1960a, b), Humbries and colleagues (1965), and Larter (1967). Chlormequat chloride was released in 1959 under the trade name Cycocel (Fletcher and Kirkwood, 1982). When wheat was grown in pots and in the field, chlormequat chloride was found to shorten the straw, whereas its efficacy in barley varied

(Tolbert, 1960b; Humbries, Welbank, and Witts, 1965; Larter, 1967). Later, individual and mixed applications of mepiquat chloride and ethephon were found to reduce stem elongation and lodging in barley more repeatedly than when chlormequat chloride was used alone (Stanca, Jenkins, and Hanson, 1979; Herbert, 1982).

The role of PGRs in global agriculture is modest when compared with that of other agrochemicals, such as fungicides, herbicides, and insecticides. Worldwide sales of PGRs are barely 4 percent of the total sales of the crop protection agents (Rademacher, 1993). Various defoliants and desiccants cover some 40 percent of the global PGR sales, followed by ethephon (23 percent), mepiquat chloride and chlormequat chloride (each 10 percent), with miscellaneous compounds accounting for the remaining 17 percent (Rademacher, 1993). Defoliants and desiccants are not used in cereals, whereas ethephon, chlormequat chloride, mepiquat chloride, and their mixtures are economically important in cereal production. The use of PGRs differs markedly from one country to another. For example, in Finland, where cereals, together with grassland, cover some 90 percent of cultivated arable land, PGRs are almost solely used to improve lodging resistance in cereals; thus, chlormequat chloride (mostly used) and ethephon, alone and in mixture with mepiquat chloride, account for 99 percent of the total PGR sales (Hynninen and Blomqvist, 1996). The sale of PGRs in Finland is, however, 4 percent of the total global sales of agrochemicals (Hynninen and Blomqvist, 1996).

HOW DO PLANT GROWTH REGULATORS WORK?

Excluding ethephon, PGRs used as antilodging agents in cereals inhibit gibberellin biosynthesis at different stages of the pathway (Adams et al., 1992; Graebe et al., 1992; Rademacher et al., 1992) (see Figure 2.1). Gibberellins stimulate stem elongation through enhancing cell elongation rather than cell division (Goodwin and Mercer, 1988). The most dramatic effects of exogenously applied gibberellins on stimulation of stem elongation are recorded for cultivars carrying dwarfing genes. Such cultivars either accumulate fewer gibberellins or are less responsive to them. For example, the presence of dwarfing genes *Rht1*, *Rht2*, and *Rht3* in wheat results in insensitivity of stem elongation to gibberellin (Gale and Youssefian, 1985). How-

FIGURE 2.1. Main Phases in the Biosynthetic Pathway of Gibberellins and the Specific Steps Blocked by Antigibberellin Compounds

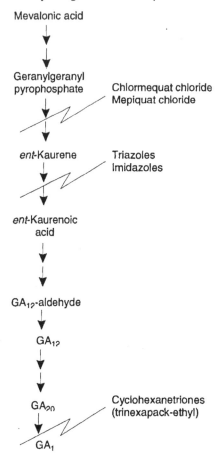

Source: Burden et al., 1987; Rademacher et al., 1987; Adams et al., 1992; Graebe et al., 1992; and Sponsel, 1995.

ever, no marked stimulation of stem elongation in naturally tall plants was recorded following exogenous gibberellin application (Goodwin and Mercer, 1988). Commonly, exogenously applied phytohormone does not result in a similar response in plants as that produced when it is endogenously synthesized.

Blocking gibberellin biosynthesis has evidently been a successful tool for reducing plant height with PGRs. The gibberellin biosynthesis pathway has four main phases: (1) the pathway from mevalonic acid to geranylgeranyl pyrophosphate (GGPP), (2) cyclization of GGPP to *ent*-kaurene, (3) *ent*-kaurene to GA_{12}-aldehyde, and (4) conversion of GA_{12}-aldehyde to other gibberellins (Sponsel, 1995). Different anti-gibberellins block the gibberellin biosynthesis pathway at different stages (see Figure 2.1). For example, onium-type compounds, such as chlormequat chloride and mepiquat chloride, block the A activity of *ent*-kaurene synthesis (Goodwin and Mercer, 1988; Graebe et al., 1992; Sponsel, 1995), whereas nitrogencontaining heterocycles, including triazoles (paclobutrazol) and imidazoles, interfere with oxidation of *ent*-kaurene to *ent*-kaurenoic acid (Burden et al., 1987; Rademacher et al., 1987; Sponsel, 1995). More recently, developed PGRs, such as cyclo-hexanetriones, including cimectacarbs (trinexapack-ethyl), interfere with gibberellin biosynthesis in the late stages of the metabolic pathway (see Figure 2.1), by reducing GA_1- induced shoot growth through inhibition of 3ß-hydroxylation of GA_{20} to GA_1 (Adams et al., 1992).

Ethephon does not, however, interfere with the gibberellin biosynthesis pathway but releases ethylene when absorbed in the cell. Ethylene is a well-known growth-retarding hormone that, for example, stimulates fruit ripening and leaf abscission, inhibits stem elongation and root growth, stimulates fading of some flowers, and inhibits formation of root nodules (Ligero, Lluch, and Olivares, 1987; Goodwin and Mercer, 1988; Peters and Crist-Estes, 1989; Lee and LaRue, 1992). Decomposition of ethephon is pH dependent; ethephon is stable at $\leq$ pH 4, but when absorbed into more neutral cells/cytoplasm, the breakdown of the compound to ethylene, chloride, and phosphate ions begins (Luckwill, 1981).

POSSIBILITIES FOR MANIPULATING FORMATION AND REALIZATION OF YIELD POTENTIAL

Yield potential is given various meanings. Therefore, the authors would like to clarify the terms *yield potential* and *potential yield,* as there is a difference between them. Potential yield is the maximum theoretical yield in tons per hectare ($t \cdot ha^{-1}$) in particular growing regions, whereas yield potential describes the way in which this theoretical maximum is obtained. Yield potential describes "the maximum storage capacity" and can be seen as the sum of its components; thresholds do exist for number of heads per square meter, number of spikelets per head, number of fertile florets and grains per spikelet and, thus, per head, and, furthermore, grain size, when account is taken of growing season length that represents the frame for production. Potential yield, often a target yield, cannot exceed these limitations.

The estimated maximum theoretical grain yield of small-grain cereals—wheat, for example—is some 18,000 kilograms per hectare ($kg \cdot ha^{-1}$), according to Gilland (1985). However, this has already been exceeded, for example, in the wheat experiments of Salisbury and Bugbee (1988, ref. Evans 1993), in which CO_2 enrichment was used. These experiments simulated future changes in atmospheric CO_2 concentration. Did they, however, just manage to reach the upper limit? Were the growing conditions comparable to those in the field in general? Was the sample size large enough for yield estimates to be taken? Or should we simply inflate the yield potential? Evans (1993) criticized the comparison of actual yields with theoretical maximum yields, saying that potential yields are often misused, representing hopes that can never be fulfilled. Evidently, we cannot fully exploit the yield potential with realistic crop management (i.e., not extremely high nitrogen fertilizer application rates, pesticide application rates, etc.), as record yields are often obtained with far from environmentally sound methods. As can be seen in Figure 2.2, the larger the growing area from which the average is calculated, the wider the gap between the estimated and recorded yields. In fact, the way we have defined the term yield potential allows us to say that there are different yield potentials in different environments and not one "universally applicable parameter." Although Figure 2.2 may appear to be a wish list, it does give some ideas about the physiological and environmental limitations to cereal yields.

Plant growth regulators that are targeted to further shorten cereal straw may enhance realization of yield potential by improving partitioning of dry matter into harvestable yield. This is due not only to reduced assimilate demand for stem elongation, and the consequent surplus of assimilates that can be used for floret and spikelet set and grain growth, but also to the direct and/or indirect effects of PGRs on

FIGURE 2.2. The Maximum Theoretical Yield, Record Yields, and Some Mean Yields for Wheat

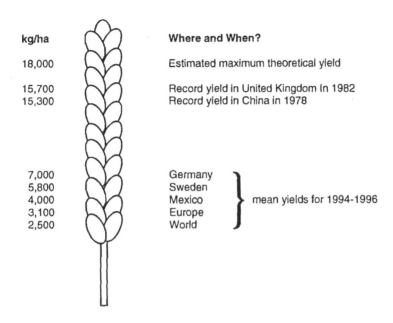

kg/ha	Where and When?
18,000	Estimated maximum theoretical yield
15,700	Record yield in United Kingdom In 1982
15,300	Record yield in China in 1978
7,000	Germany
5,800	Sweden
4,000	Mexico } mean yields for 1994-1996
3,100	Europe
2,500	World

Sources: Cheng, Bao, and Cheng, 1979, ref. Evans, 1993; Trow-Smith, 1982, ref. Evans, 1993; Gilland, 1985; and Food and Agriculture Organization, 1996b.

several plant stand characteristics that depend on timing of the application relative to crop growth stage.

MODIFYING ROOT GROWTH AND SEEDLING ESTABLISHMENT

During the course of cereal breeding, root systems have tended to become concentrated in the uppermost layers of the soil profile, thereby enhancing nutrient uptake, but making plant stands more sensitive to drought and more dependent on irrigation. Mac Key (1988) showed that most of the root mass (53 percent) in an old oat cultivar occurred below 60 centimeters (cm), whereas only 6 percent of the root mass in a modern short-strawed cultivar penetrated to such a depth. In addition to this, Siddique and colleagues (1990) reported modern wheat cultivars to have progressively lower root:shoot ratios at anthesis. Also, wild *Hordeum* species had higher root:shoot ratios than cultivated ones in the studies of Chapin and colleagues (1989). These examples suggest that PGRs may be useful in manipulation of root growth. The classical PGRs used for shortening the straw, however, are not necessarily the best candidates, even though they are occasionally advertised as root growth enhancers, without any hard evidence, as in Finland. The empirical data from PGR effects on root growth in cereals reflect their use as antilodging agents, and in this case, most often, the aim has been to monitor whether PGRs have any disadvantageous side effects on root growth.

Application of ethephon alone and mixed with mepiquat chloride, either as a seed treatment or foliar application, inhibited root growth in barley when measured two to three weeks after application (Woodward and Marshall, 1987, 1988). Similar results were recorded recently by Rajala and Peltonen-Sainio (unpublished data) with barley, oat, and wheat when treated with chlormequat chloride, ethephon, and trinexapack-ethyl. Steen and Wünsche (1990) found no clear trend for chlormequat chloride effects on root weight and root length in barley and wheat when measured posttreatment, nor did Banowetz (1993) for ethephon-treated wheat seed. Contrary results can also be found. For example, Humbries, Welbank, and Witts (1965) and Naylor and colleagues (1986, 1989) found that chlormequat chloride increased root weight in wheat and barley. Furthermore, De and colleagues (1982) recorded that root growth was stimulated in dryland wheat when seeds were treated with chlormequat chloride, which resulted in improved water uptake from deeper soil layers and higher grain yield. Bragg and colleagues (1984) noted that chlormequat chloride application at the three- to four-leaf stage for winter wheat in autumn tended to slightly increase root mass when measured the following spring. Such an effect, however, was not recorded later in the growing season. Moreover, in winter barley, chlormequat chloride treatment increased root mass, especially in deep soil layers, but this was too late in the growing season to result in any yield increase (Bragg et al., 1984). Yang and Naylor (1988) soaked triticale seeds with tetcyclacis and chlormequat chloride and showed inhibited shoot growth relative to root growth, thus an increased root to shoot ratio. Similar results were also recorded for winter barley treated with chlormequat chloride before or during early tillering stage (Naylor, Saleh, and Farquharson, 1986).

Often, if PGR treatment retards root growth, the effect is only momentary and retarded growth is followed by similar or enhanced root growth compared with a control. Therefore, the difference in root biomass, root length, and/or root depth between PGR-treated and control plants seems to vanish, or treated plants may even have greater root mass (Bragg et al., 1984; Woodward and Marshall, 1987, 1988; Yang and Naylor, 1988; Steen and Wünsche, 1990). For example, in triticale, seed treatment with PGRs first restricted shoot growth, which was, however, followed by boosting of root growth (Yang and Naylor, 1988). This may indicate that at the very early growth stages, PGRs alter partitioning of photoassimilates by favoring root growth, with the surplus assimilates not used for shoot growth.

MODIFYING STRAW CHARACTERISTICS

Are Modifications Still Needed?

One of the major aims in cereal breeding has been to shorten the straw and thereby make the plant stands more lodging resistant. Shortening of the straw, and the consequent increase in harvest index (HI), has been one of the main reasons for genetically determined yield increases achieved with barley, oat, and wheat during this century (Jain and Kulshrestha, 1976; Lawes, 1977; Riggs et al., 1981; Sinha et al., 1981; Syme and Thompson, 1981; Wych and Stuthman, 1983; Hucl and Baker, 1987; Martiniello et al., 1987; Siddique, Kirby, and Perry, 1989; Slafer and Andrade, 1989; Peltonen-Sainio, 1990; Lynch and Frey, 1993). Evans (1993) demonstrated that the gain in grain yield approximately equals the loss in straw weight, and changes in total aboveground biomass are rarely associated with yield increases. However, increase in total aboveground biomass, as documented by Waddington and colleagues (1986, 1987), Austin, Ford, and Morgan (1989), and Peltonen-Sainio (1991), is likely to have an impact on lodging sensitivity and, hence, the need for PGRs.

Despite worldwide success in breeding shorter-strawed cultivars, use of heavier management inputs—especially until the end of the

1980s—has often resulted in more dense, and thereby lodging-sensitive, plant stands. This has at least partly hidden the breeders' achievements. For example, in Finland, oat breeders shortened the straw by 20 percent between 1935 and the end of the 1980s, but the consequent 60 percent increase in lodging resistance was reduced to a 14 percent net gain due to the use of higher rates of nitrogen (N) fertilizer applications (120 kg N/ha^{-1}) that were possible for these short-strawed cultivars (Peltonen-Sainio, Granqvist, and Säynäjärvi, 1993). This example indicates that improvements at the level of genotype have not been followed by such marked changes in the phenotype; thus, opportunity exists to further modify the straw characteristics with PGRs.

Lodging interferes with yield formation in cereals, especially by causing disharmony in source-sink interaction (see Figure 2.3). If the disharmony is only transient and the plant stand rises again, the dry matter losses may remain negligible. However, in the case of severe lodging, no compensation can occur later in the growing season, and yield losses will result. The canopy structure and its ability to capture photosynthetically active radiation is far from optimal in lodged plant stands. This results in reduced photosynthate production. Severe lodging also hinders xylem and phloem transport, resulting in reduced water, nutrient, and assimilate transport (Pinthus, 1973). This results in reduced assimilate flow for grain filling. The duration of the decreased grain-filling rate depends on whether the plant stands are able to rise again. Furthermore, a lodged plant stand dries more slowly, causing delayed harvesting (see Figure 2.3), which is especially critical in growing conditions such as those in northern Europe, where rains often occur during the harvest period (Kivi, 1961). High moisture levels in the canopy favor the spread of certain saprophytic fungi that may reduce grain quality and its acceptability to the processing industry. Species and cultivars with weak seed dormancy may sprout, especially when lodged, that is, already in intact heads.

This results in severe losses in grain quality and quantity (Kivi, 1961) and is especially critical for bread wheat and malting barley, but also when high-quality seed material is produced in all cereal species. In addition to this, and especially in northern growing conditions, lodging complicates mechanical harvesting and increases demand for grain drying, thereby increasing production costs (see Figure 2.3). A thorough review of the causes and effects of lodging are given by Pinthus (1973).

FIGURE 2.3. Some Disadvantageous Effects of Lodging That Can Be Reduced or Eliminated with Antilodging Agents

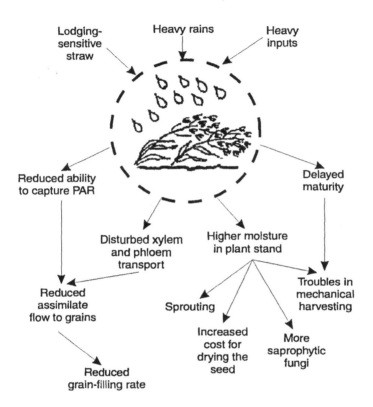

PGR EFFECTS ON STEM ELONGATION AND LODGING RESISTANCE

A noteworthy amount of literature documents that PGR-induced shortening of straw is followed by advantageous effects on yield formation of cereals in the presence and absence of lodging (Tolbert, 1960b; Humbries, Welbank, and Witts, 1965; Stanca, Jenkins, and Hanson, 1979; De et al., 1982; Naylor and Saleh, 1987; Naylor, 1989; Ma and Smith, 1991; Börner and Meinel, 1993). A summary of the effects of PGRs on straw length, lodging resistance, grain yield, and other characteristics of the cereal stand that are recorded according to different publications are shown in Table 2.1. In general, these experiments showed that PGRs shortened the straw up to 40 percent, decreased lodging to varying degrees (not very well documented), and resulted in –40 percent to +20 percent change in grain yield, depending on growing conditions and cereal species and cultivars.

Often PGRs are said not only to shorten but also to stiffen cereal straw. Introducing short-statured and dwarf cultivars, however, has not necessarily been associated with increased straw stiffness. Furthermore, contrary to this commonly held interpretation, results from the studies of Clark and Fedak (1977) and Knapp and colleagues (1987) indicated that chlormequat chloride and ethephon increased the concentrations of water-soluble carbohydrates but did not affect the concentrations of lignin and structural carbohydrates such as cellulose and hemicellulose in wheat culms. Therefore, the "mode of action" of these PGRs on decreasing lodging seems to be solely through shortening the straw rather than strengthening it through increased amounts of structural components in the cell wall (Knapp, Harms, and Volenec, 1987). This idea is also supported by the finding that stem diameter did not differ between control and anti-gibberellin-treated plants, according to Gendy and Höfner (1989), which is, however, contrary to the findings of Tolbert (1960b).

PGR EFFECTS ON BUILDUP OF STEM RESERVES

Chlormequat chloride and ethephon treatments increased the total amount of water-soluble carbohydrates in wheat culm in the experiment of Knapp and colleagues (1987), indicating that PGRs may enhance the accumulation of reserved assimilates in stems (Knapp, Harms, and Volenec, 1987; Ma and Smith, 1992a). In numerous experiments, mobilization of stem reserves—comprising excess products of photosynthesis prior to grain filling (Bidinger, Musgrave, and Fisher, 1977; Blum et al., 1991)—has been found to contribute greatly to grain yield (up to 74 percent, depending on estimation method), especially when stresses such as high temperature and drought occurred during grain growth (Gallagher, Biscoe, and Hunter, 1976; Austin et al., 1977, 1980; Bidinger, Musgrave, and Fisher,

TABLE 2.1. Summary of PGR Effects on Stem Characteristics, Grain Yield, and Other Documented Morphophysiological Traits in Small-Grain Cereals

PGR*	Cereal Species	PGR Effect Compared to Control (%)				Reference
		Plant Height	Lodging	Grain Yield	Other Traits	
CCC	Wheat	Decrease			+ tillering + Stem diam.	Tolbert (1960b)
CCC	Wheat	-40%		+5%	+ Ears/m² + grains/ear	Humbries, Welbank, and Witts (1965)
CCC	Barley	-10%-20%		No effect	+ maturity	Larter (1967)
CCC	Wheat Barley Oat	-up to 30% +5% to -15% -0%-15%		No effect No effect No effect		Clark and Fedak (1977)
ETH+ MEP	Barley	-10%	Decrease	+0% to 10%		Stanca, Jenkins, and Hanson (1979)
CCC	Wheat			+10% to 20%	+ root growth	De et al. (1982)
CCC	Wheat Barley	Decrease		No effect	+ root growth	Bragg et al. (1984)
CCC	Barley	-20%		No effect	+ leaf longevity	Green (1985)
CCC	Winter Barley				+ tillering + number of leaves/ plants - shoot:root	Naylor, Saleh, and Farquharson (1986)
MEP	Barley	Increase Decrease		Increase Decrease	+ eats/plant	Waddington and Cartwright (1986)
CCC ETH	Wheat	Decrease	Decrease	+ water-soluble carbohydrates in culm		Knapp, Harms, and Volenec (1987)

TABLE 2.1 *(continued)*

| PGR* | Cereal Species | PGR Effect Compared to Control (%) | | | | |
		Plant Height	Lodging	Grain Yield	Other Traits	Reference
CCC	Barley			No effect	+ tillering + ears/m^2	Naylor, Stokes, and Matthews (1987)
ETH+MEP	Barley			No effect	+ tillering + ear-bearing tillers − root growth	Woodward and Marshall (1987)
ETH	Wheat Barley	Decrease	Decrease	−13% to +12% −9% to +13%	− grains/ear − grain weight	Simmons et al. (1988)
ETH ETH+MEP	Barley				+ tillering − root growth + tillering	Woodward and Marshall (1988)
CCC TET	Triticale				− shoot growth − shoot:root + tillering + root growth	Young and Naylor (1988)
TRIA	Winter Wheat				− tillering temporarily − shoot growth in autumn + winter survival	Anderson (1989)
CCC ETH	Wheat	No effect No effect	No effect Decrease	No effect No effect		Cox and Otis (1989)
CCC	Wheat	Decrease			+ tillering	Craufurd and Cartwright (1989)

		PGR Effect Compared to Control (%)				
PGR*	Cereal Species	Plant Height	Lodging	Grain Yield	Other Traits	Reference
CCC	Triticale	Decrease	Decrease	Increase / No effect	+ grains/ear / − grain weight / + maturity	Naylor (1989)
ETH	Barley			Increase	+ LAD / + tillering / + ears/plant / + maturity	Ramos et al. (1989)
CCC	Triticale				+ tillers / + leaves/main shoot / + leaf lamina area / + shoot dry weight	Naylor, Brereton, and Munro (1989)
CCC	Barley	No effect		No effect	No effect on root growth	Steen and Wünsche (1990)
	Wheat	− up to 20%		No effect	+ root growth	
CCC ETH	Barley				− abortion of spikelet primordia	Ma and Smith (1991)
ETH	Barley			Decrease / Increase	+ ears/plant / − grains/ear / + maturity	Moes and Stobbe (1991)
ETH	Barley	−11% - 41%	Reduce	+3% to −64%	+ straw yield / + tillering / − grains/ear	Taylor, Foster, and Caldwell (1991)
ETH	Wheat	−10%	Reduce	+5%	+ ears/m^2	Khan and Spilde (1992)

TABLE 2.1 (continued)

PGR*	Cereal Species	Plant Height	Lodging	Grain Yield	Other Traits	Reference
			PGR Effect Compared to Control (%)			
CCC	Barley			Increase	+ dry matter in leaf blades, sheats, and chaff, stem	Ma and Smith (1992a)
ETH				Decrease Increase	grains/ear − grain mass + dry matter in leaf blades, sheats, and chaff, stem	
CCC ETH	Barley	Slight decrease − up to 25%	No effect Reduce	+ Up to 15% −40% to +10%	+ grains/ear + ears/m² − grains/ear + grains/ear	Ma and Smith (1992b)
ETH	Barley	Decrease	Decrease	No effect	− grains/ear	Stobbe et al. (1992)
ETH	Wheat			No effect	+ development rate	Banowetz (1993)
CCC	Wheat	−4% − 12%		+0% to 12%	+ grains/ear	Börner and Meinel (1993)
ETH	Barley	− up to 15%	Decrease	No marked effect	− grains/ear + maturity	Foster and Taylor (1993)
CCC	Triticale				+ grains/ear − grain weight	Naylor and Stephen (1993)
CCC ETH	Barley				+ ear weight + grains/ear − grain weight	Ma, Dwyer, and Smith (1994)
CCC ETH	Wheat				No effects	

PGR Effect Compared to Control (%)						
PGR*	Cereal Species	Plant Height	Lodging	Grain Yield	Other Traits	Reference
CCC ETH ETH+ MEP	Barley	– up to 10%	Decrease	No effect	+ maturity	Erviö et al. (1995)
CCC	Wheat			Reduce to +14%	+ head-bearing tillers	Peltonen and Peltonen-Sainio (1997), and Peltonen-Sainio and Peltonen (1997)
	Barley			Reduce	No effect	
	Oat			+7% to 12%	+ grains/head	

Notes: *CCC, chlormequat chloride; ETH, ethephon; MEP, mepiquat chloride; TRIA, triadimenol; TET, Tetcyclacis. Plus (+) and minus (–) indicate increase and decrease for the specific traits, respectively.

1977; Blum et al., 1983, 1989, 1991, 1997; Blum, 1996; Ehdaie and Waines, 1996). Blum and colleagues (1991) suggested that reserved carbohydrates may also enhance yield formation in nonstressed conditions. These additional reserves may back up the photosynthate production by better providing a continuous photoassimilate flow to the grain, when grains are sinks and demand more photosynthate than foliage is able to produce, that is, when source limitation occurs.

The crucial question is how the cereal plant utilizes the savings resulting from straw shortening, as grains are not the only sinks. Gent and Kiyomoto (1998) suggested that semidwarf wheat cultivars allocated the additional carbohydrates, resulting from less competition during stem elongation, for yield formation, as semidwarf cultivars had more fertile florets at anthesis and greater harvest index at maturity. However, is there a difference in the use of these savings if they are achieved by, for example, applying PGRs or using dwarfing genes? Use of reserve carbohydrate most likely differs in these cases—not least because activity and strength of different sinks vary during crops' growth cycles, thus making the timing of the PGR application crucial in relation to crop growth stage.

Postanthesis reserves might be even more important than preanthesis reserves for formation of yield structures (Kühbauch and Thome, 1989), making them a better potential target for manipulation with PGRs. However, it is relevant to ask whether PGRs have any economically feasible possibilities, since, for example, Austin and colleagues (1980) concluded that higher yielding wheat cultivars are less dependent on reserves, and Mehrhoff and Kühbauch (1990) reported that fructan remobilization was faster and more complete in the new wheat cultivars compared with old ones, especially when the grain-filling period was shorter.

In the experiments of Austin and colleagues (1977) and Lauer and Simmons (1988), carbohydrates stored during preanthesis in main stem and tillers contributed only modestly to grain yield. Furthermore, Rawson and Evans (1971), studying wheat, and Austin and colleagues (1980), barley, did not find preanthesis reserves of dwarf cultivars contributing more to grain yield than those of tall ones. Dwarf cultivars, in particular, may support tiller growth with the photoassimilates not needed for stem elongation, as increased tillering is very typical of dwarf cultivars (Gale and Youssefian, 1985; Peltonen-Sainio and Järvinen, 1995; Mäkelä, Väärälä, and Peltonen-Sainio, 1996). These results do not give much hope for early application of PGRs to enhance grain fill as a result of formation and mobilization of abundant carbohydrate reserves. In the experiments of Ma and Smith (1992a), chlormequat chloride and ethephon application tended to increase the weight of chaff, culm, and the three uppermost leaves in barley, which was monitored several times during the postanthesis phase. Such increased mass of nonyield structures in PGR-treated plants, however, was not associated with improved translocation of carbohydrate reserves to grain growth, but the weight of these nonyield structures was reduced more in control than in PGR-treated plants, during the grain-filling period, and resulted in variable PGR effects on grain yield (Ma and Smith, 1992a). In the experiments of Froment and McDonald (1997), CCC application followed by ethephon reduced stem length and lodging in rye. Reduced stem length was associated with reduction in stem weights, which was speculated to lessen the stress tolerance and consequently also the grain yield in rye, especially in droughted, and hence low lodging pressure, growing conditions (Froment and McDonald, 1997).

MODIFYING TILLERING AND NUMBER OF HEADS PER UNIT AREA

Do PGRs Have Any Role Left in Control of Tillering?

The number of grain-bearing heads per unit area is, in addition to the number of florets per spikelet and spikelets per head, the yield determining factor. Production of vegetative and head-bearing tillers is largely dependent on species, genotype, seeding rate, photoperiod, temperature, and water and nutrient status during the tillering period (Langer, 1972; Batten, 1985; Wright and Hughes, 1987; Peltonen-Sainio and Järvinen, 1995; Peltonen-Sainio, 1999). Tillering is an important adaptive characteristic of grasses that enables plants to utilize fully available space and resources. Ability of cereals to tiller improves their plasticity, that is, ability to respond to fluctuations in plant density (Hutley-Bull and Schwabe, 1982; Peltonen-Sainio and Järvinen, 1995). Relatively low seeding rates are commonly used in the main cereal production areas, allowing tillers to contribute to grain yield production. In northern growing conditions, seeding rates as high as 500 to 700 seeds per meter squared (seeds/m^{-2}) are favored to inhibit tillering and promote a uniculm growth habit. This is because under long days, as in northern Europe during the growing season, tillering is inhibited through stronger apical dominance (Peltonen-Sainio and Järvinen, 1995). Control of apical dominance is via endogenous plant hormone balance, especially that of auxins, cytokinins, and possible ethylene (Langer, Prasad, and Laude, 1973; Harrison and Kaufman, 1980, 1982; Salisbury and Marinos, 1985; Li and Bangerth, 1992; Tamas, 1995). Under short-day conditions more tillers are formed, but initiation rate of tiller primordia is slower when compared with initiation under long-day conditions (Hutley-Bull and Schwabe, 1982; Batten, 1985; Craufurd and Cartwright, 1989).

This brief introduction leads us to ask whether there is any remaining role for PGRs, if such an intrinsic plasticity mechanism already exists, if this characteristic is not excessively stifled during domestication of cereals (excluding uniculm genotypes), and, furthermore, if tiller performance is even easier to manipulate with, for example, seeding rate, nitrogen fertilizer rate, and watering. The results of several studies carried out with cereals show clear examples of situations in which PGRs improve productivity via effects on tillering.

PGR EFFECTS
ON TILLER GROWTH AND YIELD

Chlormequat chloride, ethephon, singly and mixed, stimulated tillering and most often resulted in more head-bearing tillers per unit area when applied as seed treatment (Woodward and Marshall, 1987) or sprayed onto foliage (Tolbert, 1960b; Humbries, Welbank, and Witts, 1965; Waddington and Cartwright, 1986; Naylor, Stokes, and Matthews, 1987; Woodward and Marshall, 1988; Craufurd and Cartwright, 1989; Ramos et al., 1989; Moes and Stobbe, 1991; Khan and Spilde, 1992; Peltonen and Peltonen-Sainio, 1997). This enhanced tillering, however, was not found to be followed by an increase in grain yield (see Table 2.1) in the studies of Tolbert (1960b), Waddington and Cartwright (1986), Woodward and Marshall (1987), Moes and Stobbe (1991), and Peltonen and Peltonen-Sainio (1997). Excessive formation of tillers, especially non-head-bearing tillers, is often considered a waste of photoassimilates, but they may also serve for temporary storage of photo assimilates during the period of excess photoassimilate production (Austin et al., 1977; Lauer and Simmons, 1988). As these examples show, tillering itself may have either a positive or a negative impact on yield formation, depending on genotype, environment, and their interaction.

Following early application of antigibberellins, such as chlormequat chloride, response of wheat to photoperiod changed, and treatment with antigibberellins mimicked the effects of short days; that is, development rate was slowed down and tillering was increased (Hutley-Bull and Schwabe, 1982; Craufurd and Cartwright, 1989). Exogenous gibberellic acid (GA_3) treatment had an opposite effect: it promoted spikelet initiation rate and development in barley and wheat (Cottrell, Dale, and Jeffcoat, 1982; Hutley-Bull and Schwabe, 1982). When barley and triticale seeds were soaked in different chlormequat chloride concentrations, tillering was markedly boosted and more leaves were produced, resulting in greater leaf lamina area and a higher shoot dry weight, which may indicate increased yield potential (Naylor, Brereton, and Munro, 1989). This was also the case with winter barley treated with chlormequat chloride before or during early tillering, as the number of emerged leaves and tillers increased (Naylor, Saleh, and Farquharson, 1986). Increased tillering was also responsible for the higher number of leaves per plant following early application of ethephon and sulphur to barley rather than due to increased number of leaves per tiller or single leaf size (Ramos et al., 1989).

Retarded elongation of the main stem following use of PGRs decreases demand for assimilates and nutrients for main shoot growth and, thereby, possibly provides excess substrates for tiller buds to be released from apical dominance and elongate (Woodward and Marshall, 1987, 1988; Ma and Smith, 1991). Due to the PGRs hormonal or antihormonal nature, changes in tillering may also be due to changes in hormonal stimulus. For example, ethephon-containing plant growth retardants release ethylene to plant tissue, thereby increasing internal ethylene concentrations. In oat, ethylene suppressed growth of the main shoot and promoted tiller bud release and growth (Harrison and Kaufman, 1982). Such a contrary effect of ethephon on growth of main shoot and tillers may be attributed to the finding that ethylene inhibits auxin biosynthesis and auxin movement in stem tissue (Evans, 1984). This could reduce the auxin flow from the apical meristem to basal regions and, hence, release tiller buds from dormancy (Woodward and Marshall, 1987).

MODIFYING FLORET, GRAIN, AND SPIKELET SET

Cereals tend to produce surplus florets and spikelets when compared with the number of filled grains and spikelets (Peltonen-Sainio and Peltonen, 1995; Peltonen and Nissilä, 1996). This indicates incomplete realization of yield potential and provides an interesting possibility for PGRs to boost the grain and spikelet set and, thereby, yield formation. According to Peltonen-Sainio and Peltonen (1995), reduction in spikelet number of spring wheat and oat occurred close to booting stage at the macromorphological scale—or, more accurately, at the apical development stage when stigmatic branches differentiate as swollen cells on styles (Q-stage), (Peltonen-Sainio and Pekkala, 1993; Peltonen-Sainio, 1999). Thus, competition for assimilates, nutrients, and so forth, among yield-determining and nonyield components coincides with initiation of spikelet abortion (Peltonen-Sainio and Peltonen, 1995). Sink capacity of these generative organs might be strengthened, and thereby frequency of abortion decreased, by maintaining the growth-stimulating phytohormone concentrations at adequate levels or at favored balance (Michael and Beringer, 1980; Peltonen-Sainio, 1997). This was reached in the experiments of Peltonen-Sainio (1997), in which exogenous synthetic cytokinin, benzylaminopurine, was applied before onset of floret abortion. In these experiments, benzylaminopurine treatment did not reduce the number of aborted florets and spikelets, but increased the total number of spikelets in an oat panicle. No marked growth-enhancing effect was recorded, as additional spikelets initiated at the lowermost branches of the axis were mostly sterile (Peltonen-Sainio, 1997). Similar experiments were also later carried out with gibberellic acid and auxin, but they also failed to prevent floret and spikelet abortion (Peltonen-Sainio and Rajala, unpublished data). Ma and Smith (1991), however, reported that abortion of spikelet primordia was inhibited when barley was sprayed with chlormequat chloride or ethephon.

Other PGRs also have been tested for increasing grain and/or spikelet number in the main shoot and tiller head. In the 1960s, Humbries, Welbank, and Witts (1965) reported that chlormequat chloride treatment resulted in more grains per head and was associated with slightly increased yield. Chlormequat chloride and ethephon applications given by peduncle perfusion during grain filling tended to increase head weight, with more grains per head and lower single grain weight, whereas no such effect was recorded in wheat (Ma, Dwyer, and Smith, 1994). Moreover, foliar applications of chlormequat chloride, ethephon, and a mixture of the two close to double-ridge stage increased grain number per head in barley, oat, and wheat (Waddington and Cartwright, 1986; Ma and Smith, 1991; Peltonen and Peltonen-Sainio, 1997). Similarly, the number of grains per ear was higher due to chlormequat chloride treatment in triticale, wheat, and barley in the experiments of Naylor (1989), Börner and Meinel (1993), and Ma and Smith (1992b), respectively. Such observations are either due to increased number of initiated spikelets and florets or decreased frequency of spikelet and floret abortion (Rahman and Wilson, 1997; Hutley-Bull and Schwabe, 1982; Peltonen and PeltonenSainio, 1997). Contrary to these examples, Moes and Stobbe (1991), Ma and Smith (1992b), Stobbe and colleagues (1992), and Foster and Taylor (1993) found that number of grains per head was reduced in barley when ethephon was sprayed at late growth stages, in particular. As for tillering, the effects of PGRs on grain and spikelet set seem to be especially dependent on time of application relative to crop growth stage, and environment. The interaction of PGR treatments, especially ethephon, with stresses is particularly disadvantageous for yield formation.

MODIFYING GRAIN-FILLING RATE

The sink capacity of the grain at grain filling is especially dependent on the number of endosperm cells, that is, cell division, which occurs during the two-week period after anthesis (Brocklehurst, 1977). Brocklehurst (1977) and Radley (1978) found a strong association between cell number and grain size in wheat. Furthermore, Michael and Seiler-Kelbitsch (1972) found two barley cultivars that differed in grain size to also differ in cytokinin activity. The postanthesis cytokinin activity was higher in the large-grained barley cultivar. Thus, we could speculate that cytokinin is one of the major phytohormones determining the potential size of the grain. After cell division, the degree to which this potential is realized depends largely on environmental factors during grain filling, such as nutrient status of the soil, water availability, temperature, and so on.

Even though a relationship seems to exist between endogenous plant hormone concentrations and potential grain size, manipulation of grain storage capacity with exogenous PGRs—chlormequat chloride, synthetic cytokinins, gibberellic acid, and ethephon, among others—seems to be rather ineffective (Radley, 1978; Ma, Dwyer, and Smith, 1994; Peltonen-Sainio, 1997). In fact, none of the publications listed in Table 2.1 showed any clear increase in grain weight following PGR treatment. Furthermore, application of chlormequat chloride by peduncle perfusion tended to decrease single grain weight in barley in the experiments of Ma, Dwyer, and Smith (1994). This, however, probably resulted from a PGR-induced increase in grain number per head, as grain yield was unaffected (Ma, Dwyer, and Smith, 1994). This is in accordance with the findings of Naylor (1989) and Naylor and Stephen (1993) in triticale. Such increased rate of head filling may result from improved leaf area duration at grain filling, as recorded in barley in the experiments of Ramos and colleagues (1989). They found that a treated plant stand reached the maximum leaf area index later than the untreated one, and this was followed by more persistent leaf area index at postanthesis.

MODIFYING DURATION OF PHENOPHASES AND SENESCENCE PROCESSES

Generally, for cereals, the shorter the duration of the period for leaf and spikelet primordia initiation, the higher the development rate of these primordia and the fewer the resulting vegetative and generative plant organs (Rahman and Wilson, 1977; Cottrell, Dale, and Jeffcoat, 1982; Hutley-Bull and Schwabe, 1982; Peltonen-Sainio, 1994; Slafer and Rawson, 1994). Effects of PGRs on duration of certain phenostages or phenophases has not been thoroughly studied. Too often, the effects on days to maturity is the only characteristic reported (see Table 2.1). Some reports on this phenomenon, however, do exist. For example, wheat seed treatment with ethephon resulted in earlier passage from a vegetative to a generative growth phase (Banowetz, 1993), which might be expected to correlate with fewer tillers. These treatments, however, did not greatly affect formation of yield components, as only the number of grains per head was slightly increased (Banowetz, 1993). Early applications of gibberellic acid under short-day conditions increased leaf and spikelet initiation rate and shortened the initiation period, that is, affected cereal development as if it had been grown under long-day conditions (Cottrell, Dale, and Jeffcoat, 1982; Hutley-Bull and Schwabe, 1982). On the other hand, application of antigibberellins had an effect similar to short days, slowing down the initiation rate of leaves and spikelets and lengthening the initiation period (Hutley-Bull and Schwabe, 1982). Furthermore, dressing seed of winter wheat with chlormequat chloride and triazole-type growth retardants reduced tillering and suppressed the growth of subcrown internodes in autumn (Anderson, 1989; Aufhammer and Federolf, 1992). Reduced elongation of belowground internodes resulted in better overwintering of the shoot apices. Seed treatment of winter and spring wheat with chlormequat chloride, to improve overwintering and drought tolerance, was practiced extensively in the former Soviet Union (Hoffmann, 1992)—the present-day situation is not known.

Plant growth regulators often tend to retard development—at least briefly. In some studies, delayed development has also been recorded as a delayed maturation (Tolbert, 1960b; Humbries, Welbank, and Witts, 1965; Larter, 1967; Moes and Stobbe, 1991; Ma and Smith, 1992b; Foster and Taylor, 1993; Erviö et al., 1995). Green (1985) showed also that chlormequat chloride delayed senescence of the lower leaves of a barley canopy. In the experiments of Peltonen- Sainio (1997), foliar application of synthetic cytokinin close to booting did not affect chlorophyll content of the flag leaf nor number of days to maturity.

POTENTIAL OF UNCONVENTIONAL PLANT GROWTH REGULATORY COMPOUNDS FOR ENHANCING YIELD FORMATION

The potential of alcohols and glycinebetaine as growth-enhancing agrochemicals was studied largely in the 1990s; hence, their possible growth-enhancing properties are briefly speculated on in the following sections.

Foliar Application of Alcohols

Nonomura and Benson (1992) applied only methanol or nutrient-supplemented methanol to the foliage of numerous C_3 crops under warm, arid growing conditions of Arizona and established marked increases in water-use efficiency and biomass accumulation. There are several suggestions for the mode of action of applied methanol. One possibility is that methanol is utilized through the photorespiratory pathway; a nonbeneficial pathway is used to detoxify methanol, and lost photosynthates are returned for use by the plant (Haugstad et al., 1983; Nonomura and Benson, 1992). In pea (*Pisum sativum* L.), alcohols increased thylakoid-bound fructose-1,6-bisphosphatase, which is one of the principal enzymes controlling the activity of the photosynthetic carbon reduction cycle (Andrés et al., 1990). Such activity was dependent on the concentration of applied alcohol as well as on the length of the carbon chain of the alcohol (Andrés et al., 1990). In addition to this, alcohols have delayed senescence through inhibiting ethylene effects (Heins, 1980; Satler and Thimann, 1980; Saltveit, 1989). Delayed senescence is likely to prolong the period of high photosynthetic activity in leaves, increase carbon dioxide (CO_2) fixation, and thereby result in better realization of yield potential—especially if this is the case in the uppermost leaves of the canopy during the postanthesis phase.

The article of Nonomura and Benson (1992) created great interest in the use of methanol and other alcohols as crop growth enhancers. Devlin, Bhowmik, and Karczmarczyck (1994) and Rowe, Farr, and Richards (1994) reported improved seedling growth in greenhouse-grown tomato (*Lycopersicon esculentum* Mill.), pea, radish (*Raphanus sativus* L.), and wheat following foliar application of alcohol. However, the majority of the results from recent experiments indicate no growth-enhancing effect in cereals or other crop species (Hartz et al., 1994; McGiffen et al., 1994; Wutcher, 1994; Albrecht et al., 1995; Esensee, Leskovar, and Boales, 1995; Feibert et al., 1995; van Iersel et al., 1995; Rajala et al., 1998). In light of these results, alcohol applications seem to have, at best, limited potential as growth enhancers.

Exogenously Applied Glycinebetaine

Glycinebetaine is synthesized in the chloroplast stroma by two-step oxidation of amino acid choline. It is an osmoprotectant that is accumulated in certain plant species, halophytes in particular, when exposed to stresses, including salt and drought (McCue and Hanson, 1990; Papageorgiou, Jufimura, and Murata, 1991; Murata et al., 1992). Even though many important crop species do not accumulate this stress-induced quaternary ammonium compound, smallgrain cereals are able to synthesize it (Wyn Jones and Storey, 1981).

Contrary to the case of traditional PGRs, no clear evidence suggests that exogenously applied glycinebetaine interacts with endogenous phytohormones. No changes in abscisic acid (ABA) concentration were noted after foliar application of glycinebetaine, even though stomatal conductance increased in the studies of Mäkelä et al. (1998b). They speculated that some interaction with cytokinins may occur. Thus, the mode of action of glycinebetaine is through osmotic adjustment (Wyn Jones and Storey, 1981), protection of the oxygen (O_2)-evolving machinery of chloroplasts (Papageorgiou, Jufimura, and Murata, 1991; Murata et al., 1992; Mamedow, Hyashi, and Murata, 1993), and protection of leaf cell plasma membranes and chloroplast thylakoid membranes (Yang, Rhodes, and Joly, 1996) rather than interfering with hormone metabolism. The mode of action of exogenously applied glycinebetaine may differ from that of endogenously synthesized glycinebetaine, as is also the case with plant hormone application.

The possibilities for improving stress tolerance and increasing productivity through foliar application of glycinebetaine have been studied intensively in several crop species (Itai and Paleg, 1982; Mäkelä, Mantila, et al., 1996; Mäkelä, Peltonen-Sainio, et al., 1996; Agboma, Jones, Peltonen-Sainio, and Pehu, 1997; Agboma, Jones, Peltonen-Sainio, et al., 1997; Agboma, Peltonen-Sainio, et al., 1997; Agboma, Sinclair, et al., 1997; Mäkelä, 1998; Mäkelä et al., 1997; Mäkelä, Jokinen, et al., 1998; Mäkelä et al., 1999; Mäkelä et al., 1998a). Mäkelä, Peltonen-Sainio, and colleagues (1996) showed that when glycinebetaine was sprayed onto plant foliage as an aqueous solution, it was readily taken up by wheat, and surfactants further enhanced the uptake. Furthermore, when ^{14}C-labeled glycinebetaine was applied to turnip rape (*Brassica rapa* spp. *oleifera* L.) leaves, it began to be translocated almost immediately—during the first hours to roots and later to actively growing plant parts. It is a xylem-phloem-mobile compound (Mäkelä, Peltonen-Sainio, et al., 1996). Therefore, foliar application proved to be a feasible method. Another possibility would be to genetically engineer a crop to synthesize glycinebetaine or to enhance the existing ability to accumulate glycinebetaine (McCue and Hanson, 1990). As a nitrogenous compound, however, glycinebetaine is energetically expensive to plants, and such transgenic plants might be characterized by having a lower yield potential. Storey, Ahmad, and Wyn Jones (1977) indicated that in halophytes even 10 to 28 percent of leaf nitrogen was sequestered as glycinebetaine.

In addition to the ability of the crop to take up and translocate sprayed glycinebetaine, any growth-enhancing effect should last long enough to result in yield increases. Mäkelä, Peltonen-Sainio, and colleagues (1996) showed that glycinebetaine remained unmetabolized up to seventeen days after application. However, dilution of glycinebetaine is likely to occur as biomass increases (Agboma, Peltonen-Sainio, et al., 1997).

Under drought and salt stress, foliar application of glycinebetaine most often tends to result in yield enhancement, but when no marked stresses occur application may even result in yield losses. Such a negative yield response is possibly due to retarded growth following glycinebetaine-induced physiological changes that mimic the situation when plants are exposed to stresses. Some evidence (Mäkelä, 1998) also suggests that foliar application of glycinebetaine is not necessarily followed by enhanced growth of yield components; instead, growth of vegetative organs is stimulated. Furthermore, small-grain cereals—possibly due to their ability to synthesize glycinebetaine—are not as likely to be potential targets for glycinebetaine sprayings as dicotyledon species that do not endogenously accumulate glycinebetaine (see Table 2.2). These examples suggest source-sink interaction in relation to growth stage, induction of stress, and timing of glycinebetaine application to have great impact on the feasibility of glycinebetaine treatment as a growth enhancer. Further studies are needed, however, to thoroughly understand these interactions and, on this basis, to better address the glycinebetaine treatments to result in improved stress tolerance and higher yields.

SOME FUTURE PROSPECTS

According to some indications, the use of PGRs for cereals is not going to increase on a global scale in the near future. Awareness of the adverse effects of, for example, using high fertilizer inputs hand in hand with numerous agrochemicals, PGRs included, has stimulated a "post-green revolution" movement toward sustainable agriculture. This, together with modern breeding methods that develop day by day, has brought into question the need for PGRs in cereal production. Introducing dwarfing genes into, for example, wheat (Syme, 1970; Gent and Kiyomoto, 1998), barley (Ali, Okiror, and Rasmussen, 1978), and oat (Brown, McKenzie, and Mikaelsen, 1980; Meyers, Simmons, and Stuthman, 1985; Mäkelä, Väärälä, and Peltonen-Sainio, 1996) has enabled marked shortening and stiffening of the straw, often with the additional advantage of having more dry matter partitioned into the economic yield, that is, higher harvest index. In fact, the green revolution, on the lines laid down by Norman Borlaug, was based on the use of dwarf cultivars (Chrispeels and Sadava, 1994). In addition to use of more lodging-resistant cultivars, we are today trying to prevent nutrient, especially nitrogen and phosphorus, leaching into the environment by using fertilizer application rates that, if possible, equal the take-up rate of the cereal stand. Realization of these aspirations is directed, for example, through governmental subsidy policy. In fact, in Europe, total use of nitrogenous fertilizers in the 1990s has dropped to the level of the early 1970s, after having peaked at the end of the 1980s (Food and Agriculture Organization, 1996a). Of course, numerous factors other than environmental concerns are behind this trend, but it indicates the direction. Furthermore, consumers' timeless concerns about pesticide residues in cereal products add pressure to reduce their use, if possible—even though many modern agrochemicals are used at very low concentrations, are readily degraded in the plant and soil, and, as new products must pass through very strict health and safety examinations and tests, according to requirements of legislative authorities.

For these reasons, and possibly many others, it is likely that fewer PGRs will be used in cereals in the future, unless any "revolutionary" innovations with PGRs targeted to cereal production are forthcoming. During the forty-year history of PGRs, they have served only as antilodging agents, whereas in horticulture, PGRs have much more versatile uses than merely reducing plant height. These are, for example, induction of flowering, reducing fruit number, altering the form of fruit clusters (e.g., grapes), modifying the shape of fruits (e.g., apple), delaying maturity, synchronizing the ripening of fruits, facilitating harvesting, and much more (Gianfagna, 1995). These

TABLE 2.2. Summary of Glycinebetaine Effects on Yield Compared to Non-Betaine-Treated Control at Its Highest

Crop Species	Applied Glycinebetaine Concentration	Growth Stage at Application	Treatment	Effect on Yield Compared with Control	Measured Yield Parameter	Reference
Barley	25 mM	Young seedlings	Water stress	≤ 3%	Shoot dry weight	Itai and Paleg (1982)
Barley	0, 3.5, 10.5, 17.5 kg/ha	4-leaf stage	Irrigation	≤ 6%	Grain yield	Mäkelä, Mantila, et al. (1996)
			No irrigation	≤ 1%		
Wheat				≤ 1%	Grain yield	
				≤ 9%		
Turnip rape				≤11%	Seed yield	
				≤13%		
Barley	0, 1, 3 kg/ha	I 4-leaf stage		≤ 0%	Grain yield	Mäkelä, Mantila, et al. (1996)
		II floret aboration		≤ 0%		
Oat				≤ 0%	Grain yield	
				≤ 0%		
Wheat				≤ 3%	Grain yield	
				≤ 4%		
Turnip rape	0, 1, 3, 5 kg/ha	I 4-leaf stage		≤ 1%	Seed yield	
		II flowering		≤ 0%		
Sweet lupin	0, 2, 4, 6 kg/ha		Drought	≤22%	Seed yield	Agboma, Jones, Peltonen-Sainio, and Pehu (1997)
			Well watered	≤14%		
Maize	0, 2, 4, 6 kg/ha		Drought	≤22%	Grain yield	Agboma, Jones, Peltonen-Sainio, et al. (1997)
			Well watered	≤14%		
Sorghum				≤12%		
				≤11%		
Wheat				≤ 7%		
				≤22%		

Crop Species	Applied Glycinebetaine Concentration	Growth Stage at Application	Treatment	Effect on Yield Compared with Control	Measured Yield Parameter	Reference
Tobacco	0, 0.1, 0.3 M	Vegetative period	Drought	≤32%	Leaf dry weight	Agboma, Peltonen-Sainio, et al. (1997)
Soybean	0, 1, 3, 6 kg/ha	I full bloom	Severe drought Mild drought Well watered	≤ 0% ≤ 6% ≤13%	Seed yield	Agboma, Sinclair, et al. (1997)
		II pod initiation		≤ 0% ≤ 4% ≤ 0%		
Tomato	0, 1.1, 2.3, 3.4 kg/ha	Early flowering	Salt stress Heat stress	≤ 5% ≤23%	Fruit yield	Mäkelä, Jokinen, et al. (1998)
		Mid-flowering	Salt stress Heat stress	≤63% ≤56%		
		Later flowering	Salt stress Heat stress	≤12% ≤15%		
Pea	0, 1, 2, 3, 4, 6, 9, 15 kg/ha	3-leaflet stage		≤16%	Plant dry weight	Mäkelä et al. (1999)
Turnip rape		Early rosette stage		≤36%		

examples indicate that humankind can and does manipulate crop growth substantially. Cereals—due to their great importance (47 percent of total global crop production area in 1996) (FAO, 1996b) rather than being a niche crop—are likely to attract research and development. Thus, interesting new opportunities, as yet unknown, may arise tomorrow for PGRs in cereal production.

Regarding the use of traditional antilodging agents, further information is needed to fully benefit from their use. Therefore, we need to understand fully the possible adverse effects of PGRs on yield formation and use this knowledge in decision making. Before applying PGRs to cereal stands, farmers need adequate information on weather conditions, phenology of the plant, physiological stage of the plant (i.e., is it suffering from stresses?), among many other interacting factors. Farmers should also have the "gift of foresight" to determine when use of PGRs in cereals is truly needed and worthwhile.

CONCLUDING REMARKS

This chapter has summarized some of the possible means to manipulate cereal growth through application of PGRs to increase grain yield production. These ideas are represented also in Figure 2.4. The authors' impression from preparing this chapter is that only one characteristic, almost without exception, seems to be altered with PGR applications, namely plant height. Reducing plant height is the primary target for these PGR compounds. Other plant characteristics, such as root growth; tillering; floret, grain, and spikelet set; harvest index; rate of grain filling; relative length of phenophases; and days to maturity, vary greatly from one experiment to another, and conflicting results are easily obtained. This is because the effect of traditional PGRs, the ones that can be classified as antilodging agents, is greatly dependent on (1) application time in relation to crop growth stage, (2) its interaction with and effect on partitioning of current photosynthate, as well as (3) reserve and mobilized carbohydrates, and, furthermore, (4) their dependence on weather conditions, especially occurrence of stresses, and, last but not least, (5) genotypic differences in response to PGRs.

FIGURE 2.4. Possible Targets for Manipulation of Cereal Stands, Their Source-Sink Interaction, and Yield Formation with PGRs

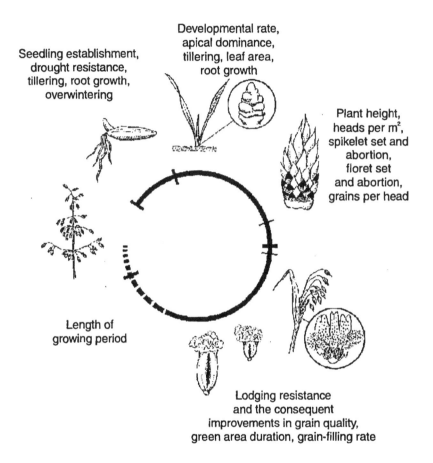

Seedling establishment, drought resistance, tillering, root growth, overwintering

Developmental rate, apical dominance, tillering, leaf area, root growth

Plant height, heads per m^2, spikelet set and abortion, floret set and abortion, grains per head

Length of growing period

Lodging resistance and the consequent improvements in grain quality, green area duration, grain-filling rate

REFERENCES

Adams, R., E. Kerber, K. Pfister, and E.W. Weiler (1992). Studies on the action of the new growth retardant CGA 163´935 (Cimectacarb). In *Progress in Plant Growth Regulation, Proceedings of the 14th International Conference on Plant Growth Substances,* C. Karssen, L. Van Loon, and D. Vreugdenhill (Eds.). Dordrecht, Netherlands: Kluwer Academic Publishers, pp. 818-827.

Agboma, P.C., M.G.K. Jones, P. Peltonen-Sainio, and E. Pehu (1997). Effects of exogenous glycinebetaine on seed yield and alkaloid content of sweet lupin under two watering regimes. *The Science of Legumes* 4:242-248.

Agboma, P.C., M.G.K. Jones, P. Peltonen-Sainio, H. Rita, and E. Pehu (1997). Exogenous glycinebetaine enhances grain yield of maize, sorghum and wheat grown under two supplementary watering regimes. *Journal of Agronomy and Crop Science* 178:29-37.

Agboma, P.C., P. Peltonen-Sainio, R. Hinkkanen, and E. Pehu (1997). Effects of foliar application of glycinebetaine on yield components of drought–stressed tobacco plants. *Experimental Agriculture* 33:345–352.

Agboma, P.C., T.R. Sinclair, K. Jokinen, P. Peltonen-Sainio, and E. Pehu (1997). An evaluation of the effect of foliar application of glycinebetaine on the growth and yield of soybean: Timing of application, watering regimes and cultivars. *Field Crops Research* 54:51-64.

Albrecht, S.L., C.L. Douglas, E.L. Klepper, P.E. Rasmussen, R.W. Rickman, R.W. Smiley, D.E. Wilkins, and D.J. Wysocki (1995). Effects of foliar methanol applications on crop yield. *Crop Science* 35:1642-1646.

Ali, M.A.M., S.O. Okiror, and D.C. Rasmusson (1978). Performance of semidwarf barley. *Crop Science* 18:418-422.

Anderson, H.M. (1989). Effects of triadimenol seed dressing on vegetative growth in winter wheat. *Crop Research* 29:29-36.

Andrés, A.R., J.J. Làzaro, A. Chueca, R. Hermoso, and J.L. Gorgé (1990). Effect of alcohols on the association of photosynthetic fructose-1,6-bisphosphatase to thylakoid membranes. *Physiologia Plantarum* 78:409-413.

Aufhammer, W. and K.G. Federolf (1992). Plant growth regulator seed treatments for improved cold tolerance in hard wheat (*Triticum durum*). *Bodenkultur* 43:29-38.

Austin, R.B., J.A. Edrich, M.A. Ford, and R.D. Blackwell (1977). The fate of dry matter, carbohydrates and ^{14}C lost from leaves and stems of wheat during grain filling. *Annals of Botany* 41:1309-1321.

Austin, R.B., M.A. Ford, and C.L. Morgan (1989). Genetic improvement in the yield of winter wheat: A further evaluation. *Journal of Agricultural Science, Cambridge* 112:295-301.

Austin, R.B., C.L. Morgan, M.A. Ford, and R.D. Blackwell (1980). Contributions to grain yield from pre-anthesis assimilation in tall and dwarf barley phenotypes in two contrasting seasons. *Annals of Botany* 45:309-319.

Banowetz, G.M. (1993). The effect of ethephon seed treatment on leaf development and head initiation of wheat. *Field Crops Research* 34:113-120.

Batten, G.D. (1985). Oligolum wheat: Tillering and ontogeny in relation to daylength and temperature. *Cereal Research Communications* 13:97-100.

Bidinger, F.R., R.B. Musgrave, and R.A. Fisher (1977). Contribution of stored pre-anthesis assimilate to grain yield in wheat and barley. *Nature* 270:431-433.

Blum, A. (1996). The role of mobilized stem reserves in stress tolerance. *V International Oat Conference & VII International Barley Genetics Symposium,* G. Scoles and B. Rossnagel (Eds.). Saskatoon: University Extension Press, pp. 267-275.

Blum, A., G. Golan, J. Mayer, and B. Sinmena (1997). The effect of dwarfing genes of sorghum grain filling from mobilized stem reserves under stress. *Field Crops Research* 52:43-54.

Blum, A., G. Golan, J. Mayer, B. Sinmena, L. Shpiler, and J. Burra (1989). The drought response of landraces of wheat from the northern Negev Desert in Israel. *Euphytica* 43:87-96.

Blum, A., H. Poiarkova, G. Golan, and J. Mayer (1983). Chemical desiccation of wheat plants as simulator of post-anthesis stress. II. Effects on translocation and kernel growth. *Field Crops Research* 6:51-58.

Blum, A., L. Shpiler, G. Golan, J. Mayer, and B. Sinmena (1991). Mass selection of wheat for grain filling without transient photosynthesis. *Euphytica* 54:111-116.

Börner, A and A. Meinel (1993). The effects of the growth retardant chlormequat (CCC) on plant height and yield in GA insensitive wheats. *Plant Breeding* 110:255-258.

Bragg, P.L., P. Rubino, F.K.G. Henderson, W.J. Fielding, and R.Q. Cannel (1984). A comparison of the root and shoot growth of winter barley and winter wheat, and the effect of an early application of chlormequat. *Journal of Agricultural Science, Cambridge* 103:257-264.

Brocklehurst, P.A. (1977). Factors controlling grain weight in wheat. *Nature* 266:348-349.

Brown, P.D., R.I.H. McKenzie, and K. Mikaelsen (1980). Agronomic, genetic, and cytologic evaluation of a vigorous new semidwarf oat. *Crop Science* 20:303-306.

Burden, R.S., G.A. Carter, T. Clark, D.T. Cooke, S.J. Croker, A.H. Deas, P. Hedden, C.S. James, and J.R. Lenton (1987). Comparative activity of the enantiomers of triadimenol and paclobutrazol as inhibitors of fungal growth and plant sterol and gibberellin biosynthesis. *Pesticide Science* 21:253-267.

Chapin, F.S., R.H. Groves, and L.T. Evans (1989). Physiological determinants of growth rate in response to phosphorus supply in wild and cultivated *Hordeum* species. *Oecologia* 79:96-105.

Cheng, D.-Z., X.K. Bao, and Z. Cheng (1979). A preliminary study on morphological and physiological indices of high yielding spring wheat in Chaidamu Basin. *Aelea Agriculture* 2:29-39. (In Chinese with English abstract.)

Chrispeels, M.J. and D.E. Sadava (1994). *Plants, Genes, and Agriculture.* Boston, MA: Jones and Bartlett Publishers, 478 pp.

Clark, R.V. and G. Fedak (1977). Effects of chlormequat on plant height, disease development and chemical constituents of cultivars of barley, oats and wheat. *Canadian Journal of Plant Science* 57:31-36.

Cottrell, J.E., J.E. Dale, and B. Jeffcoat (1982). The effects of daylength and treatment with gibberellic acid on spikelet initation and development in Clipper barley. *Annals of Botany* 50:57-68.

Cox, W.J. and D.J. Otis (1989). Growth and yield of winter wheat as influenced by chlormequat chloride and ethephon. *Agronomy Journal* 81:264-270.

Craufurd, P.Q. and P.M. Cartwright (1989). Effect of photoperiod and chlormequat on apical development and growth in a spring wheat (*Triticum aestivum*) cultivar. *Annals of Botany* 63:515-525.

De, R., G. Giri, G. Saran, R.K. Singh, and G.S. Chaturvedi (1982). Modification of water balance of dryland wheat through the use of chlormequat chloride. *Journal of Agricultural Science, Cambridge* 98:593-597.

Devlin, R.M., P.C. Bhowmik, and S.J. Karczmarczyck (1994). Influence of methanol on plant growth. *Plant Growth Regulation Society of America Quarterly* 22:102-108.

Ehdaie, B. and J.G. Waines (1996). Genetic variation for contribution of preanthesis assimilates to grain yield in spring wheat. *Journal of Genetics and Breeding* 50:47-55.

Erviö, L.-R., H. Jalli, M. Kontturi, H. Hakkola, A. Kangas, and P. Simojoki (1995). Benefits of using plant growth regulators in fodder barley. *Agricultural Science in Finland* 4:429-443.

Esensee, V., D.I. Leskovar, and A.K. Boales (1995). Inefficacy of methanol as a growth promoter in selected vegetable crops. *Hort Technology* 5:253-256.

Evans, L.T. (1993). *Crop Evolution: Adaptation and Yield.* Cambridge: Cambridge University Press, 500 pp.

Evans, M.L. (1984). Functions of hormones at the cellular level of organization. In *Hormonal Regulation of Development. II. The Function of Hormones from the Level of the Cell to the Whole Plant. Encyclopedia of Plant Physiology,* New Series 10, T.K. Scott (Ed.). Berlin: Springer-Verlag, pp. 23-79.

Feibert, E.B.G., S.R. James, K.A. Rykbost, A.R. Mitchell, and C.C. Shock (1995). Potato yield and quality not changed by foliar-applied methanol. *HortScience* 30:494-495.

Fletcher, W.W. and R.C. Kirkwood (1982). *Herbicides and Plant Growth Regulators.* Granada Publishing, 408 pp.

Food and Agriculture Organization (FAO) (1996a). *Fertilizer Year Book,* Volume 46. Rome: FAO, 122 pp.

Food and Agriculture Organization (FAO) (1996b). *Production Year Book,* Volume 50. Rome: FAO, 235 pp.

Foster, K.R. and J.S. Taylor (1993). Response of barley to ethephon: Effects of rate, nitrogen, and irrigation. *Crop Science* 33:123-131.

Froment, M.A. and H.G. McDonald (1997). Effect of a plant growth regulator regime on internode length and weight of tillers in conventional and hybrid rye and the impact of nitrogen on crop performance. *Journal of Agricultural Science, Cambridge* 129:143-154.

Gale, M.D. and S. Youssefian (1985). Dwarfing genes in wheat. In *Progress in Plant Breeding,* Volume 1, E. Russell (Ed.). London: Butterworths, pp. 1-35.

Gallagher, J.N., P.V. Biscoe, and B. Hunter (1976). Effects of drought on grain growth. *Nature* 264:541-542.

Gendy, A. and W. Höfner (1989). Stalk shortening of oat (*Avena sativa* L.) by combined application of CCC, DCiB and ethephon. *Angew Botanik* 63:103-110.

Gent, M.P.N. and R.K. Kiyomoto (1998). Physiological and agronomic consequences of *Rht* genes in wheat. *Journal of Crop Production* 1:27-46.

Gianfagna, T.T. (1995). Natural and synthetic growth regulators and their use in horticultural and agronomic use. In *Plant Hormones and Their Role in Plant Growth and Development*, P.J. Davies (Ed.). Dordrecht, Netherlands: Kluwer Academic Publishers, pp. 751-773.

Gilland, B. (1985). Cereal yields in theory and practice. *Outlook on Agriculture* 14:56-60.

Goodwin, T.W. and Mercer, E.I. (1988). *Introduction to Plant Biochemistry.* Oxford: Pergamon Press, 677 pp.

Graebe, J.E., G. Böse, E. Grosselindemann, P. Hedden, H. Aach, A. Schweimer, and T. Lange (1992). The biosynthesis of *ent*-kaurene in germinating seeds and the function of 2-oxoglutarate in gibberellin biosynthesis. In *Progress in Plant Growth Regulation, Proceedings of the 14th International Conference on Plant Growth Substances*, C. Karssen, L. Van Loon, and D. Vreugdenhill (Eds.). Dordrecht, Netherlands: Kluwer Academic Publishers, pp. 545-554.

Green, D.G. (1985). Effects of CCC and GA on internodal development of barley. *Plant and Soil* 86:291-294.

Harrison, M.A. and P.B. Kaufman (1980). Hormonal regulation of lateral bud (tiller) release in oats (*Avena sativa* L.). *Plant Physiology* 66:1123-1127.

Harrison, M.A. and P.B. Kaufman (1982). Does ethylene play a role in the release of lateral buds (tillers) from apical dominance in oats? *Plant Physiology* 70:811-814.

Hartz, T.K., K.S. Mayberry, M.E. McGiffen, M. LeStrange, G. Miyao, and A. Baameur (1994). Foliar methanol application ineffective in tomato and melon production. *HortScience* 29:1087.

Haugstad, M., L. Ulsaker, A. Ruppel, and S. Nilsen (1983). The effect of triacontanol on growth, photosynthesis and photorespiration in *Chlamydomonas reinhardtii* and *Anacystis nidulans. Physiologia Plantarum* 58:451-456.

Heins, R.D. (1980). Inhibition of ethylene synthesis and senescence in carnation by ethanol. *Journal of American Society of Horticultural Science* 105:141-144.

Herbert, C.D. (1982). Growth regulation in cereals—Chance or design? In *Chemical Manipulation of Crop Growth and Development*, J.S. McLaren (Ed.). London: Butterworth Scientific, pp. 315-327.

Hoffmann, G. (1992). Use of plant growth regulators in arable crops: Survey and outlook. In *Progress in Plant Growth Regulation, Proceedings of the 14th International Conference on Plant Growth Substances*, C. Karssen, L. Van Loon, and D. Vreugdenhil (Eds.). Dordrecht, Netherlands: Kluwer Academic Publishers, pp. 798-808.

Hucl, P. and R.J. Baker (1987). A study of ancestral and modern Canadian spring wheats. *Canadian Journal of Plant Science* 67:87-97.

Humbries, E.C., P.J. Welbank, and K.J. Witts (1965). Effect of CCC (chlorocholine chloride) on growth and yield of spring wheat in the field. *Annals of Applied Biology* 56:351-361.

Hutley-Bull, P.D. and W.W. Schwabe (1982). Some effects of low-concentration gibberellic acid and retardant application during early growth on morphogenesis in wheat. In *Chemical Manipulation of Crop Growth and Development*, J.S. McLaren (Ed.). London: Butterworth Scientific, pp. 329-342.

Hynninen, E. and H. Blomqvist (1996). Pesticide sales in Finland in 1995. *Kemia-Kemi* 23:485-488.

Itai, C. and L.G. Paleg (1982). Responses of water-stressed *Hordeum distichum* L. and *Cucumis sativus* to proline and betaine. *Plant Science Letters* 25:329-335.

Jain, H.K. and V.P. Kulshrestha (1976). Dwarfing genes and breeding for yield in bread wheat. *Zeitschrift für Pflanzenzüchtung* 76:102-112.

Khan, A. and L. Spilde (1992). Agronomic and economic response of spring wheat cultivars to ethephon. *Agronomy Journal* 84:399-402.

Kivi, E.I. (1961). Sprouting in the head on spring wheat varieties. *Maatalous ja Koetoiminta* 15:101-109. (In Finnish with English summary.)

Knapp, J.S., C.L. Harms, and J.J. Volenec (1987). Growth regulator effects on wheat culm nonstructural and structural carbohydrates and lignin. *Crop Science* 27:1201-1205.

Kühbauch, W. and U. Thome (1989). Nonstructural carbohydrates of wheat stems as influenced by sink-source manipulations. *Journal of Plant Physiology* 134:243-250.

Langer, R.H. (1972). *How Grasses Grow*. London, Beccles, Colchester: William Clowes and Sons Ltd., 60 pp.

Langer, R.H., P.C. Prasad, and H.M. Laude (1973). Effects of kinetin on tiller bud elongation in wheat (*Triticum aestivum* L.). *Annals of Botany* 37:565-571.

Larter, E.N. (1967). The effect of (2-chlorethyl) trimethylammonium chloride (CCC) on certain agronomic traits on barley. *Canadian Journal of Plant Science* 47:413-421.

Lauer, J.G. and S.R. Simmons (1988). Photoassimilate partitioning by tillers and individual tiller leaves in field-grown spring barley. *Crop Science* 28:279-282.

Lawes, D.A. (1977). Yield improvement in spring oats. *Journal of Agricultural Science, Cambridge* 89:751-757.

Lee, K.H. and T.A. LaRue (1992). Ethylene as a possible mediator of light- and nitrate-induced inhibition of nodulation of *Pisum sativum* L. cv. Sparkle. *Plant Physiology* 100:1334-1338.

Li, C.J. and F. Bangerth (1992). The possible role of cytokinins, ethylene and indoleacetic acid in apical dominance. In *Progress in Plant Growth Regulation, Proceedings of the 14th International Conference on Plant Growth Substances*, C. Karssen, L. Van Loon, and D. Vreugdenhill (Eds.). Dordrecht, Netherlands: Kluwer Academic Publishers, pp. 431-436.

Ligero, F., C. Lluch, and J. Olivares (1987). Evolution of ethylene from roots and nodulation rate of *Alfalfa medicago-sativa* L. plants inoculated with *Rhizobium-*

meliloti as affected by the presence of nitrate. *Journal of Plant Physiology* 129: 461-466.

Luckwill, L.C. (1981). *Growth Regulators in Crop Production.* Studies in Biology No. 129. Southampton, England: The Camelot Press Ltd., 59 pp.

Lynch, P.J. and K.J. Frey (1993). Genetic improvement in agronomic and physiological traits of oat since 1914. *Crop Science* 33:984-988.

Ma, B.L. and D.L. Smith (1991). Apical development of spring barley in relation to chlormequat and ethephon. *Agronomy Journal* 83:270-274.

Ma, B.L. and D.L. Smith (1992a). Growth regulator effects on aboveground dry matter partitioning during grain fill of spring barley. *Crop Science* 32:741-746.

Ma, B.L. and D.L. Smith (1992b). Chlormequat and ethephon timing and grain production of spring barley. *Agronomy Journal* 84:934-939.

Ma, B.L., L.M. Dwyer, and D.L. Smith (1994). Evaluation of peduncle perfusion for *in vivo* studies of carbon and nitrogen distribution in cereal crops. *Crop Science* 34:1584-1588.

Mac Key, J. (1988). Shoot:root interrelations in oats. In *3rd International Oat Conference,* B. Mattsson and R. Lyhagen (Eds.). Svalöv, Sweden, pp. 340-344.

Mäkelä, P. (1998). Foliar application of glycinebetaine and plant physiological response in tomato and turnip rape. Helsinki, Finland: Department of Plant Production, University of Helsinki, publications, No. 52, 46 pp.

Mäkelä, P., K. Jokinen, M. Kontturi, P. Peltonen-Sainio, E. Pehu, and S. Somersalo (1998). Foliar application of glycinebetaine—a novel product from sugar beet— as an approach to increase tomato yield. *Industrial Crops and Products* 7:139-148.

Mäkelä, P., J. Kleemola, K. Jokinen, J. Mantila, E. Pehu, and P. Peltonen-Sainio (1997). Growth response of pea and summer turnip rape to foliar application of glycinebetaine. *Acta Agriculturae Scandinavica,* Section B, *Soil and Plant Sciences* 47:168-175.

Mäkelä, P., M. Kontturi, E. Pehu, and S. Somersalo (1999). Photosynthetic response of drought- and salt-stressed tomato and turnip rape to foliar-applied glycinebetaine. *Physiologia Plantarum* 105:45-50.

Mäkelä, P., J. Mantila, R. Hinkkanen, E. Pehu, and P. Peltonen-Sainio (1996). Effect of foliar applications of glycinebetaine on stress tolerance, growth, and yield of spring cereals and summer turnip rape in Finland. *Journal of Agronomy and Crop Science* 176:223-234.

Mäkelä, P., R. Munns, T.D. Colmer, A.G. Condon, and P. Peltonen-Sainio (1998a). Effect of foliar applications of glycinebetaine on stomatal conductance, abscisic acid and solute concentrations of salt and drought-stressed tomato. *Australian Journal of Plant Physiology* 25:655-663.

Mäkelä, P., R. Munns, T.D. Colmer, A.G. Condon, and P. Peltonen-Sainio (1998b). Effect of foliar applications of glycinebetaine on stomatal conductance, abscisic acid and solute concentrations of salt and drought-stressed tomato. *Australian Journal of Plant Physiology* 25:655-663.

Mäkelä, P., P. Peltonen-Sainio, K. Jokinen, E. Pehu, H. Setälä, R. Hinkkanen, and S. Somersalo (1996). Uptake and translocation of foliarly applied glycinebetaine in crop plants. *Plant Science* 121:221-230.

Mäkelä, P., L. Väärälä, and P. Peltonen-Sainio (1996). Agronomic comparison of Minnesota-adapted dwarf oat with semi-dwarf, intermediate, and tall oat lines adapted to northern growing conditions. *Canadian Journal of Plant Science* 76:727-734.

Mamedow, M.D., H. Hyashi, and N. Murata (1993). Effects of glycinebetaine and unsaturation of membrane lipids on heat stability of photosynthetic electron-transport and phosphorylation reactions in *Synechocystis* PCC6803. *Biochemica et Biophysica Acta* 1142:1-5.

Martiniello, P., G. Delogu, M. Odoardi, G. Boggini, and A.M. Stanca (1987). Breeding progress in grain yield and selected agronomic characters of winter barley (*Hordeum vulgare* L.) over the last quarter of a century. *Plant Breeding* 99:289-294.

McCue, K.F. and A.D. Hanson (1990). Drought and salt tolerance: Toward understanding and application. *Trends in Biotechnology* 8:358-362.

McGiffen, M.E., J. Manthey, A. Baameur, R.L. Greene, B.A. Faber, A.J. Downer, and J. Aguiar (1994). Field tests of methanol as a crop yield enhancer. *HortScience* 29:459.

Mehrhoff, R. and W. Kühbauch (1990). Yield components in old and new German winter wheat varieties with respect to the storage and remobilization of fructan in the wheat stem. *Journal of Agronomy and Crop Science* 165:47-53.

Meyers, K.B., S.R. Simmons, and D.D. Stuthman (1985). Agronomic comparison of dwarf and conventional height oat genotypes. *Crop Science* 25:964-966.

Michael, G. and H. Beringer (1980). The role of hormones in yield formation. In *Physiological Aspects of Crop Productivity, Proceedings of the 15th Colloquium of the International Potash Institute.* Bern, Switzerland, pp. 85-116.

Michael, G. and H. Seiler-Kelbitsch (1972). Cytokinin content and kernel size of barley grain as affected by environmental and genetic factors. *Crop Science* 12:162-165.

Moes, J. and E.H. Stobbe (1991). Barley treated with ethephon. I. Yield components and net grain yield. *Agronomy Journal* 83:86-90.

Murata, N., P.S. Mohanty, H. Hayashi, and G.C. Papageorgiou (1992). Glycinebetaine stabilizes the association of extrinsic proteins with the photosynthetic oxygen-evolving complexes. *FEBS Letters* 296:187-189.

Naylor, R.E.L. (1989). Effects of the plant growth regulator chlormequat on plant form and yield in triticale. *Annals of Applied Biology* 114:533-544.

Naylor, R.E.L., P.S. Brereton, and L. Munro (1989). Modification of seedling growth of triticale and barley by seed-applied chlormequat. *Plant Growth Regulation* 8:117-125.

Naylor, R.E.L. and M.E. Saleh (1987). Effects of plant spacing and chlormequat on the plant structure, growth and yield of winter barley. *Crop Research* 27:97-109.

Naylor, R.E.L., M.E. Saleh, and J.M. Farquharson (1986). The response to chlormequat of winter barley growing at different temperatures. *Crop Research* 26:17-31.

Naylor, R.E.L. and N.H. Stephen (1993). Effects of nitrogen and the plant growth regulator chlormequat on grain size, nitrogen content and amino acid composition of triticale. *Journal of Agricultural Science, Cambridge* 120:159-169.

Naylor, R.E.L., D.T. Stokes, and S. Matthews (1987). Chemical manipulation of growth and development in winter barley production systems. *Field Crops Abstracts* 40:277-289.

Nonomura, A.M. and A.A. Benson (1992). The path of carbon in photosynthesis: Improved crop yields with methanol. *Proceedings of the National Academy of Sciences, USA* 89:9794-9798.

Papageorgiou, G.C., Y. Jufimura, and N. Murata (1991). Protection of the oxygen-evolving Photosystem II complex by glycinebetaine. *Biochemica et Biophysica Acta* 1057:361-366.

Peltonen, J. and E. Nissilä (1996). Pre- and post-anthesis duration of two-rowed barleys in relation to stability of grain yield at high latitudes. *Hereditas* 124:217-222.

Peltonen, J. and P. Peltonen-Sainio (1997). Breaking uniculm growth habit of spring cereals at high latitudes by crop management. I. Leaf area index and biomass accumulation. *Journal of Agronomy and Crop Science* 178:79-86.

Peltonen-Sainio, P. (1990). Genetic improvements in the structure of oat stands in northern growing conditions during this century. *Plant Breeding* 104:340-345.

Peltonen-Sainio, P. (1991). High phytomass producing oats for cultivation in northern growing conditions. *Journal of Agronomy and Crop Science* 166:90-95.

Peltonen-Sainio, P. (1994). Growth duration and above-ground dry-matter partitioning in oats. *Agricultural Science in Finland* 3:195-198.

Peltonen-Sainio, P. (1997). Nitrogen fertilizer and foliar application of cytokinin affect spikelet and floret set and survival in oat. *Field Crops Research* 49:169-176.

Peltonen-Sainio, P. (1999). Growth and development of oat with special reference to source-sink interaction and productivity. In *Physiological Control of Growth and Yield in Field Crops*, D.L. Smith and C. Hamel (Eds.). Berlin: Heidelberg-Springer, pp. 39-66.

Peltonen-Sainio, P., M. Granqvist, and A. Säynäjärvi (1993). Yield formation in modern and old oat cultivars under high and low nitrogen regimes. *Journal of Agronomy and Crop Science* 171:268-273.

Peltonen-Sainio, P. and P. Järvinen (1995). Seeding rate effects on tillering, grain yield, and yield components of oat at high latitude. *Field Crops Research* 40:49-56.

Peltonen-Sainio, P. and T. Pekkala (1993). Numeric codes for developmental stages of oat apex in the growing conditions of southern Finland. *Agricultural Science in Finland* 2:329-336.

Peltonen-Sainio, P. and J. Peltonen (1995). Floret set and abortion in oat and wheat under high and low nitrogen regimes. *European Journal of Agronomy* 4:253-262.

Peltonen-Sainio, P. and J. Peltonen (1997). Breaking uniculm growth habit of spring cereals at high latitudes by crop management. II. Tillering. Grain yield and yield components. *Journal of Agronomy and Crop Science* 178:87-95.

Peters, N.K. and D.K. Crist-Estes (1989). Nodule formation is stimulated by the ethylene inhibitor aminoethoxyvinylglycine. *Plant Physiology* 91:690-693.

Pinthus, M.J. (1973). Lodging in wheat, barley, and oats: The phenomenon, its causes, and preventive measures. *Advances in Agronomy* 25:209-256.

Rademacher, W. (1993). PGRs—Present situation and outlook. *Acta Horticulturae* 329:296-308.

Rademacher, W., H. Fritsch, J.E. Graebe, H. Sauter, and J. Jung (1987). Tetcyclacis and triazole-type plant growth retardants: Their influence on the biosynthesis of gibberellins and other metabolic processes. *Pesticide Science* 21:241-252.

Rademacher, W., K.E. Temple-Smith, D.L. Griggs, and P. Hedden (1992). The mode of action of acyclohexanediones—A new type of growth retardant. In *Progress in Plant Growth Regulation, Proceedings of the 14th International Conference on Plant Growth Substances*, C. Karssen, L. Van Loon, and D. Vreugdenhill (Eds.). Dordrecht, Netherlands: Kluwer Academic Publishers, pp. 571-577.

Radley, M. (1978). Factors affecting grain enlargement in wheat. *Journal of Experimental Botany* 29:919-934.

Rahman, M.S. and J.H. Wilson (1977). Determination of spikelet number in wheat. I. Effect of varying photoperiod on ear development. *Australian Journal of Agricultural Research* 28:565-574.

Rajala, A., J. Kärkkäinen, J. Peltonen, and P. Peltonen-Sainio (1998). Foliar applications of alcohols failed to enhance growth and yield of C_3 crops. *Industrial Crops and Products* 7:129-137.

Ramos, J.M., L.F. Garcia del Moral, J.L. Molina-Cano, P. Salamanca, and F. Roca de Togores (1989). Effects of an early application of sulphur or ethephon as foliar spray on the growth and yield of spring barley in a Mediterranean environment. *Journal of Agronomy and Crop Science* 163:129-137.

Rawson, H.M. and L.T. Evans (1971). The contribution of stem reserves to grain development in a range of wheat cultivars of different height. *Australian Journal of Agricultural Research* 22:851-863.

Riggs, T.J., P.R. Hanson, N.D. Start, D.M. Miles, C.L. Morgan, and M.A. Ford (1981). Comparison of spring barley varieties grown in England and Wales between 1880 and 1980. *Journal of Agricultural Science, Cambridge* 97:599-610.

Rowe, R.N., D.J. Farr, and B.A.J. Richards (1994). Effects of foliar and root applications of methanol or ethanol on the growth of tomato plants (*Lycopersicon esculentum* Mill.). *New Zealand Journal of Crop and Horticultural Science* 22:335-337.

Salisbury, F.B. and B.G. Bugbee (1988). Space farming in the 21st Century. *21st Century Science & Technology* 1:32-41.

Salisbury, F.B. and N.G. Marinos (1985). The ecological role of plant growth substances. De-etiolation and plant hormones. In *Hormonal Regulation Develop-*

ment III. *Encyclopedia of Plant Physiology*, R.P. Pharis and D.M. Reid (Eds.). New Series 11, Heidelberg: Springer-Verlag, pp. 707-750.

Saltveit, M.E. (1989). Effect of alcohols and their interaction with ethylene on the ripening of epidermal pericarp discs of tomato fruit. *Plant Physiology* 90:167-174.

Satler, S.O. and K.V. Thimann (1980). The influence of aliphatic alcohols on leaf senescence. *Plant Physiology* 66:395-399.

Siddique, K.H.M., R.K. Belford, and D. Tennant (1990). Root:shoot ratios of old and modern, tall and semi–dwarf wheats in a Mediterranean environment. *Plant and Soil* 121:89–98.

Siddique, K.H.M., E.J.M. Kirby, and M.W. Perry (1989). Ear:stem ratio in old and modern wheat varieties; relationship with improvement in number of grains per ear and yield. *Field Crops Research* 21:59-78.

Simmons, S.R., E.A. Oekle, J.V. Wiersma, W.E. Lueschen, and D.D. Warnes (1988). Spring wheat and barley responses to ethephon. *Agronomy Journal* 80:829-834.

Sinha, S.K., P.K. Aggarwal, G.S. Chaturvedi, K.R. Koundal, and R. Khanna-Chopra (1981). A comparison of physiological and yield characteristics in old and new wheat varieties. *Journal of Agricultural Science, Cambridge* 97:233-236.

Slafer, G.A. and F.H. Andrade (1989). Genetic improvement in bread wheat (*Triticum aestivum*) yield in Argentina. *Field Crops Research* 21:289-296.

Slafer, G.A. and H.M. Rawson (1994). Does temperature affect final numbers of primordia in wheat? *Field Crops Research* 39:111-117.

Sponsel, V.M. (1995). The biosynthesis and metabolism of gibberellins in higher plants. In *Plant Hormones: Physiology, Biochemistry and Molecular Biology*, P.J. Davies (Ed.). Dordrecht, Netherlands: Kluwer Academic Publishers, pp. 66-97.

Stanca, A.M., G. Jenkins, and P.R. Hanson (1979). Varietal responses in spring barley to natural and artificial lodging and to growth regulator. *Journal of Agricultural Science, Cambridge* 93:449-456.

Steen, E. and U. Wünsche (1990). Root growth dynamics of barley and wheat in field trials after CCC application. *Swedish Journal of Agricultural Research* 20:57-62.

Stobbe, E.H., J. Moes, R.W. Bahry, R. Visser, and A. Iverson (1992). Environment, cultivar, and ethephon rate interactions in barley. *Agronomy Journal* 84:789-794.

Storey, R., N. Ahmad, and R.G. Wyn Jones (1977). Taxonomic and ecological aspects of the distribution of glycinebetaine and related compounds in plants. *Oecologia* 27:319-332.

Syme, J.R. (1970). A high yielding Mexican semi-dwarf wheat and the relationship of yield to harvest index and other varietal characteristics. *Australian Journal of Experimental Agriculture and Animal Husbandry* 10:350-353.

Syme, J.R. and J.P. Thompson (1981). Phenotypic relationships among Australian and Mexican wheat cultivars. *Euphytica* 30:467-481.

Tamas, I.E. (1995). Hormonal regulation of apical dominance. In *Plant Hormones: Physiology, Biochemistry and Molecular Biology*, Second Edition, P.J. Davies (Ed.). Dordrecht, Netherlands: Kluwer Academic Publishers, pp. 572-597.

Taylor, J.S., K.R. Foster, and C.D. Caldwell (1991). Ethephon effects on barley in central Alberta. *Canadian Journal of Plant Science* 71:983-995.

Tolbert, N.E. (1960a). (2-chlorethyl)trimethylammonium chloride and related compounds as plant growth substances. I. Chemical structure and bioassay. *The Journal of Biological Chemistry* 235:475-479.

Tolbert, N.E. (1960b). (2-chloroethyl)trimethylammonium chloride and related compounds as plant growth substances. II. Effect on growth of wheat. *Plant Physiology* 35:380-385.

Trow-Smith, R. (1982). Hants wheat yield beats world record. *Farmers Weekly*, August 20, 1982.

van Iersel, M.V., J.J. Heitholt, R. Wells, and D.M. Oosterhuis (1995). Foliar methanol applications to cotton in the southeastern United States: Leaf physiology, growth, and yield components. *Agronomy Journal* 87:1157-1160.

Waddington, S.R. and P. Cartwright (1986). Modification of yield components and stem length in spring barley by the application of growth retardants prior to main shoot stem elongation. *Journal of Agricultural Science, Cambridge* 107:367-375.

Waddington, S.R., M. Osmanzai, M. Yoshida, and J.K. Ransom (1987). The yield of durum wheats released in Mexico between 1960 and 1984. *Journal of Agricultural Science, Cambridge* 108:469-477.

Waddington, S.R., J.K. Ransom, M. Osmanzai, and D.A. Saunders (1986). Improvement in the yield potential of bread wheat adapted to Northwest Mexico. *Crop Science* 26:698-703.

Woodward, E.J. and C. Marshall (1987). Effects of seed treatment with plant growth regulator on growth and tillering in spring barley (*Hordeum distichum* cv. Triumph). *Annals of Applied Biology* 110:629-638.

Woodward, E.J. and C. Marshall (1988). Effect of plant growth regulators and nutrient supply on the tiller bud outgrowth in barley (*Hordeum distichum* L.). *Annals of Botany* 61:347-354.

Wright, D. and L.L. Hughes (1987). Relationship between time, temperature, daylength and development in spring barley. *Journal of Agricultural Science, Cambridge* 109:365-373.

Wutcher, H. (1994). The effect of methanol on orange fruit quality. *HortScience* 29:575.

Wych, R.D. and D.D. Stuthman (1983). Genetic improvement in Minnesota-adapted oat cultivars released since 1923. *Crop Science* 3:879-881.

Wyn Jones, R.G. and R. Storey (1981). In *The Physiology and Biochemistry of Drought Resistance in Plants*, L.G. Paleg and D. Aspinall (Eds.). Sydney, Australia: Academic Press, pp. 171-204.

Yang, G., D. Rhodes, and R.J. Joly (1996). Effects of high temperature on membrane stability and chlorophyll fluorescence in glycinebetaine-deficient and glycinebetaine-containing maize lines. *Australian Journal of Plant Physiology* 23:437-443.

Yang, W. and R.E.L. Naylor (1988). Effect of tetcyclacis and chlormequat applied to seed on seedling growth of triticale cv. Lasko. *Plant Growth Regulation* 7:289-301.

Chapter 3

Regulation of Gibberellins
Is Crucial for Plant Stress Protection

R. Austin Fletcher
Coralie R. Sopher
Nataraj N. Vettakkorumakankav

INTRODUCTION

Plant growth and development is regulated by an interaction between hormones, including gibberellins (GAs), auxins, cytokinins, abscisic acid (ABA), and ethylene. The elusive flowering hormone "florigen," jasmonic acid, brassinosteroids, and other yet undiscovered compounds may be added to this list in the future. Unlike animal hormones, which have specific effects, plant hormones work in balance; hence, it is inappropriate to discuss them in isolation. In many instances, they have a sequential role and their ultimate effect would be dependent on the dynamic equilibrium attained by these hormones at a specific stage of plant growth and development. However, in spite of this limitation, this chapter will be a discussion on GA and its role in plant stress protection.

Work in Japan led to the discovery of GA when Kurosawa, a plant pathologist, noted that the bakanae (foolish seedling) disease of rice was caused by a fungus. The sexual form of this fungus is *Gibberella fujikuroi,* and the asexual stage is *Fusarium moniliforme.* In 1926, Kurosawa demonstrated that the medium in which the fungus was cultured contained a substance that stimulated growth of rice, and in 1935, Yabuta identified this substance and named it gibberellin (loc cit, Stowe and Yamaki, 1957). Besides their effect on elongation, it soon became apparent that GAs have several roles, including breaking of seed and bud dormancy; induction of amylase during germination; and promotion of flowering in some photoperiodically sensitive and cold-requiring plants. In general, GA was considered to stimulate the hydrolytic and degradative enzymes and, consequently, to accelerate leaf senescence. Although this effect was the case with several species, in 1965, Fletcher and Osborne unequivocally demonstrated that GA retards senescence of dandelion leaves, and since then GA has been found to retard senescence in other species, such as dock and nasturtium. These observations further confirm the difficulty of ascribing a specific effect to a single hormone. The effectiveness of exogenous application of hormones depends on the species, physiological age, as well as concentration and time of application. In addition to this problem in investigating hormone action, an understanding of GA is further complicated by the existence of more than one hundred GAs, with distinct functions suggested for some of them.

Gibberellins are diterpenes synthesized by the isoprenoid pathway (see Figure 3.1), which also generates many other terpenoids with hormonal functions, including cytokinins and abscisic acid. This pathway also produces other important metabolites: sterols, which are localized in plant membranes; carotenoids, which have a protective function against photodamage; the phytol side chain for the effectiveness of the chlorophyll molecule; and antioxidants such as α-tocopherol (Goodwin and Mercer, 1990). Furthermore, animal, fungal, and insect hormones are also products of the isoprenoid pathway. This suggests an evolutionary significance and an interdependence between organisms; for example, insects depend totally on plants for their cholesterol requirements. We propose that growth, differentiation, development, and senescence are regulated by preferential production of specific compounds of the pathway during the organism's life cycle. Several agrochemicals and pharmaceuticals have been synthesized to regulate various steps in this pathway (see Figure 3.1). Examples of these are herbicides, insecticides, fungicides, plant growth regulators (PGRs), and contraceptives that target carotenoids, insect juvenile hormone, ergosterol, GA, and estrogen production, respectively.

INHIBITORS OF GA BIOSYNTHESIS

Numerous plant growth retardants have been developed for the inhibition of plant growth. Some of these compounds inhibit the syn-

FIGURE 3.1. The Isoprenoid Pathway Is Derived from a Five-Carbon (C-5) Isoprene Unit and Generates Animal (a), Fungal (f), Plant (p), and Insect (i) Hormones

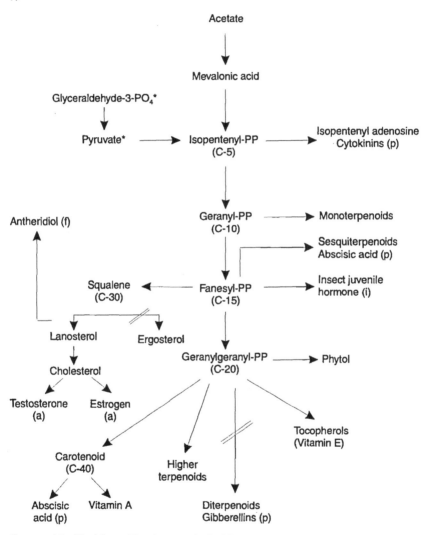

Source: Modified from Fletcher et al., 2000, p. 69.

Notes: *Alternative route to the isoprenoid pathway as proposed by Hedden and Kamiya (1997). Triazoles inhibit (//) ergosterol and gibberellins in fungi and plants, respectively.

thesis of *ent*-kaurene, an early step in GA biosynthesis, which is catalyzed by the enzyme *ent*-kaurene synthase. These early products, including (2-chloroethyl)trimethyl ammonium chloride (CCC), 2,4-dichlorobenzyl-tributylphosphonium chloride (Phosphon), ammonium (5-hydroxycarvacryl)trimethyl chloride piperidine carboxylate (AMO-1618), and their applications in agriculture, have been reviewed (Cathey, 1964; Weaver, 1972). In certain cases, the growth-inhibitory effects of these compounds were also associated with plant stress protection (Nickell, 1982). Cytochrome (Cyt) P-450-mediated reactions are inhibited by the triazoles, which affects the conversions of *ent*-kaurene to *ent*-kaurenoic acid in GA biosynthesis and lanosterol to ergosterol in fungi. Later steps in the biosynthesis of GA are catalyzed by 2-oxoglutarate-dependent dioxygenases that are inhibited by a new generation of PGRs, the cyclohexadiones (Rademacher, 1997) (see Figure 3.2). Among the inhibitors of GA biosynthesis, the triazoles are more potent as plant growth retardants and stress protectants. Hence, in this chapter, we will discuss in detail the PGR and stress-protective properties of the triazoles.

THE TRIAZOLES

The triazoles are the largest and most important group of systemic compounds, developed in the 1960s for the control of fungal diseases in plants and animals. Commercial triazole derivatives (see Table 3.1) have been recommended for use as either fungicides or PGRs; however, in varying degrees, they exhibit both properties (Fletcher, Hofstra, and Gao, 1986). They contain a common chlorophenyl group, a carbon side chain, and a 1,2,4-triazol group, with a lone pair of electrons on the sp2-hybridized nitrogen atom. The heterocyclic nitrogen atom of the triazole binds to the protoheme iron atom in Cyt P-450 systems, thereby excluding oxygen. This inhibits the conversion of lanosterol to ergosterol in fungi and *ent*-kaurene to *ent*-kaurenoic acid, a precursor of GA in plants (Rademacher, 1997). This accounts for their fungitoxic and PGR properties, and their relative activity is dependent on the stereochemical configuration of the substituents on the carbon chain (Fletcher and Hofstra, 1988). It has been reported that an R configuration at the chiral carbon bearing the

FIGURE 3.2. Simplified Pathway of Gibberellin Biosynthesis, Illustrating Reactions Catalyzed by Monooxygenases (M) and Dioxygenases (D)

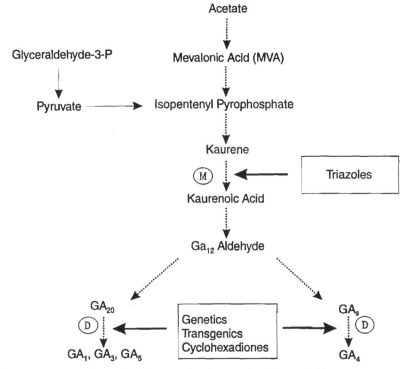

Source: Adapted from Vettakkorumakankav et al., 1999, p. 543.

Note: Sites of inhibition by chemicals (e.g., triazoles and cyclohexadiones), as well as genetic and transgenic technologies, are shown.

TABLE 3.1. Representative Examples of Triazole Compounds Recommended for Use As Fungicides or Plant Growth Regulators

Common Name	Trade Name	Application	Source
Diconazole	Spotless	Fungicide	Sumitomo
Paclobutrazol	Clipper	Plant Growth Regulator	Zeneca
Propiconazole	Tilt	Fungicide	Ciba-Geigy
Triadimefon	Bayleton	Fungicide	Bayer
Triadimenol	Baytan	Fungicide	Bayer
Uniconazole	Sumagic	Plant Growth Regulator	Sumitomo

hydroxyl group has fungicidal activity, whereas an S configuration at this carbon exhibits activity as a PGR (Fletcher et al., 2000). Compared to other PGRs, triazoles are effective at relatively low doses and are nonphytotoxic (Davis, Steffens, and Sankhla, 1988). It has been proposed (Fletcher and Hofstra, 1985) that the broad spectrum PGR properties of the triazoles are mediated through an alteration in the balance of plant hormones, including GA, ABA, and cytokinins, whereas auxin levels are not affected. In addition to their action as fungicides and PGRs, triazoles increase the tolerance of various monocot and dicot species, including conifers, to biotic and abiotic stresses, such as fungal pathogens, drought, air pollutants, and low and high temperatures. Hence, they have been referred to as "plant multiprotectants" (Fletcher and Hofstra, 1985).

Triazoles and Hormonal Changes

Inhibition of GA biosynthesis is the primary plant growth regulatory effect of the triazoles, and the relevance of this inhibition to plant stress protection will be discussed later. A transient increase in ABA levels following triazole treatment in bean plants was first noted by Asare-Boamah and colleagues (1986), and similar observations in cell suspensions, excised leaves, and whole seedlings have been reported (Grossmann, 1992). It has been suggested that increased ABA levels in triazole-treated plants were associated with prevention of its catabolism to phaseic acid, an enzymatic reaction that is catalyzed by a Cyt P-450-dependent monooxygenase (see Figure 3.3) (Rademacher, 1997). However, Häuser, Jung, and Grossmann (1992) observed a decline in ABA levels in treated plants and linked it to catabolism and/or inhibition of ABA biosynthesis. Conflicting reports on the effects of triazoles on ABA levels might be attributed to the time of analysis after triazole treatment, concentration of triazole applied, and species and developmental stage of the plant (Mackay et al., 1990; Häuser, Jung, and Grossmann, 1992).

Senescence has been delayed by triazoles in several plant species, and this has been associated with increased cytokinin levels (Fletcher et al., 2000). Although it had been reported that triazoles have cytokinin-like activity with antisenescence properties (Buchenauer and Rohner, 1981), subsequent studies showed that triazoles are not active as cytokinins but induce plants to produce more cytokinins (Fletcher and Arnold, 1986). Triazole-treated plants are typically

FIGURE 3.3. Proposed Involvement of Different Isoforms of Cyt P-450-Dependent Monooxygenases in GA and ABA Metabolism

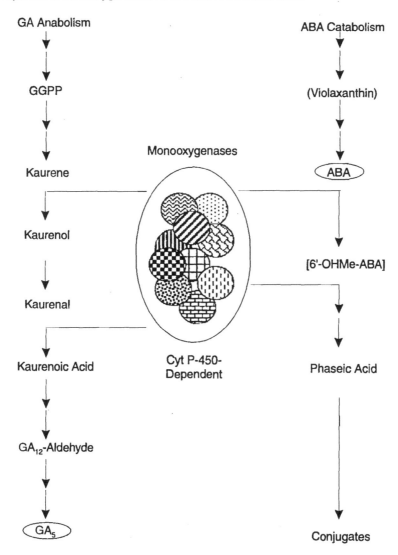

Source: Adapted from Rademacher, 1997, p. 30.

Note: GGPP = geranylgeranyl pyrophosphate.

darker green and have higher levels of chlorophyll and carotenoids, characteristic of higher cytokinin levels (Fletcher and Hofstra, 1988).

Triazoles inhibit ethylene biosynthesis in a wide range of species (Fletcher et al., 2000). Following heat stress in wheat, or the application of the auxinic herbicide triclopyr to soybeans, there was a higher accumulation of 1-aminocyclopropane carboxylic acid (ACC) in the triazole-treated seedlings. From these results, it has been suggested that triazoles inhibit the conversion of ACC to ethylene by the ethylene-forming enzyme (EFE) (Kraus, Murr, and Fletcher, 1991). Subsequent studies with EFE suggested that Cyt P-450 monooxygenase reactions could be involved in the conversion of ACC to ethylene (Kraus et al., 1992). Inhibition of ethylene production in triazole-treated sunflower cell suspensions was accompanied by a concomitant increase in ACC and malonyl (M)-ACC, and Grossmann (1992) suggested that triazoles inhibit ACC oxidase. From these studies, it can be concluded that triazoles inhibit the enzyme that converts ACC to ethylene. Delayed senescence in oilseed rape and soybean cotyledons by triazoles has been associated with decreased ethylene production (Fletcher et al., 2000). In addition to ethylene, Grossmann and colleagues (1994) observed a reduction in ABA and an increase in cytokinins. This supports the proposal by Fletcher and Hofstra (1985) that triazole-induced effects are mediated by a change in the balance of plant hormones.

Triazoles and Morphological Changes

Characteristic effects of triazoles on plants include reduced height and stem width, along with increased compactness, the extent of which is dependent on plant species, age, as well as dose and method of application (Davis, Steffens, and Sankhla, 1988). Reduced height is a consequence of triazole-induced GA inhibition, exemplified by reduced internodal elongation. Shorter stems have been correlated with decreased cell number, short cortical cells, and reduced xylem length (Fletcher et al., 1999). This compacting effect of the triazoles has been exploited commercially for many horticultural plants, trees and grasses (reviewed by Davis and Curry, 1991). Triazoles stimulate or inhibit root formation, depending on the plant species and concentration of chemical applied; and at stimulatory concentrations, they increase root:shoot ratio (Fletcher and Hofstra, 1988).

Triazole-treated plants characteristically have smaller leaves, but they are wider and thicker with more cuticular wax than controls (Gao, Hofstra and Fletcher, 1988). The leaves, therefore, have increased leaf dry weight per unit area (Davis, Steffens, and Sankhla, 1988). Increased leaf thickness has been correlated with increased cell depth and diameter and/or additional cell layers (Gao, Hofstra, and Fletcher, 1988; Burrows, Boag, and Steward, 1992). Leaves from triazole-treated plants have been reported to exhibit altered orientation (Davis, Steffens, and Sankhla, 1988), with partially closed or sunken stomata (Gao, Hofstra, and Fletcher, 1988). Light-scattering spectroscopy and microscopy has established that the cross-sectional areas of triazole-treated chloroplasts are significantly larger than those in untreated leaves (Fletcher et al., 1999). An increase in cytokinins by triazoles could lead to the observed enhanced chloroplast size and chlorophyll levels. In maize, paclobutrazol (PBZ) treatment did not change the number of chloroplasts, but there was more chlorophyll per chloroplast. The treatment increased stromal lamellae and reduced the number of grana stacks in mesophyll chloroplasts (Sopher et al., 1999).

Triazoles and Stress Protection

Triazole-treated plants characteristically use less water, have increased tolerance to drought and a higher water potential than controls (Davis, Steffens, and Sankhla, 1988; Fletcher and Hofstra 1988). Increased drought resistance in wheat seedlings was associated with reduced transpiration caused by decreased leaf area and increased wax production (Gao, Hofstra, and Fletcher, 1988). Increased diffusive resistance indicating partial closure of stomata could be caused by the observed transient rise in ABA levels (Asare-Boamah et al., 1986). Under conditions of water deficiency, triadimefon increased yield of peas and soybeans (Fletcher and Nath, 1984) and fresh weight of tomato (Fletcher and Hofstra, 1985). PBZ has been shown to induce drought resistance in conifers (Marshall, Scarratt, and Dumbroff, 1991) and this effect, along with reduced water usage by treated plants, has been exploited commercially in products such as "Confer." In wheat seedlings, PBZ treatment increased rooting and reduced the loss of membrane integrity and photosynthetic efficiency associated with waterlogging in the untreated controls (Webb and Fletcher, 1996). It has been suggested

(Mackay et al., 1990) that protection from water stress by triazoles may, in part, be the result of their effects on increasing the concentrations of ABA and amino acids, specifically proline.

Triazoles increase the tolerance of several plant species to chilling and freezing temperatures. Enhanced chilling tolerance in triazole-treated cucumber (Upadhyaya et al., 1989) and tomato (Senaratna et al., 1988) was associated with increased antioxidant enzyme concentrations. In treated tomatoes, besides the increase in the antioxidants α-tocopherol and ascorbate, free fatty acids were higher and there was a reduction in the loss of membrane phospholipids as compared to the untreated controls. Triazole-induced tolerance to low temperature stress has been associated with increased levels of endogenous ABA (Fletcher et al., 1999), which has been reported to trigger the genetic processes for hardening (Zeevaart and Creelman, 1988). In field studies, winter survival of peas and cereal crops (reviewed by Davis, Steffens, and Sankhla, 1988) and resistance to frost damage in corn and tomatoes (Fletcher and Kraus, 1995) were enhanced by triazoles.

Triazoles have been reported to increase the tolerance of plants to high temperature stress. It was suggested (Fletcher and Hofstra, 1988) that this increased thermotolerance was related to changes in the hormonal balance which could harden the plants to subsequent stress. In wheat, uniconazole-increased thermotolerance was associated with lowering of leaf temperature through increased evapotranspiration (Booker et al., 1991). Exposure of wheat seedlings to 50°C for 5 h caused thermal injury and induced several heat shock proteins (HSPs) in the controls, but not in the thermotolerant PBZ-treated plants. It was concluded that HSPs did not play a significant role in protection of treated seedlings from thermal injury (Kraus, Pauls and Fletcher, 1995). PBZ-induced thermal protection of wheat was associated with increased levels of ascorbate, glutathione, ascorbate peroxidase, SOD, guiacol peroxidase and catalase (Kraus and Fletcher, 1994). This suggests that an increased ability of treated plants to scavenge free radicals plays a significant role in triazole-induced thermotolerance.

Protection from SO_2 and O_3 damage by triazoles has been observed in several plant species and it has been proposed that this protection is mediated, in part, through decreased stomatal aperture and an increase in lipid-soluble antioxidants (Fletcher et al., 1999). Triazoles have been reported to increase tolerance of soybean seedlings to destructive levels of UV-B radiation and this protection was correlated with triazole-induced leaf thickening, enhanced cuticular wax deposition and increased levels of SOD and catalase (Kraus et al., 1995). The hypothesis that triazole-induced protection of plants from several environmental stresses including water, low and high temperatures, and air pollutants, is mediated by an increase in antioxidant potential was confirmed by demonstrating that leaves from triazole-treated plants are protected from the herbicide paraquat, a free radical generator (Kraus, McKersie, and Fletcher, 1995). From field studies conducted both in Canada and India it is evident that the degree of protection by triazoles is greater under hostile environmental conditions, especially with less tolerant cultivars. From these observations we concluded that plants have an intrinsic stress protective mechanism and triazoles allow this potential to be expressed (Fletcher and Kraus, 1995).

Although the triazoles are capable of protecting plants from a variety of biotic and abiotic stresses, their use has been limited due to their persistence in the environment. To address this concern, a novel seed treatment procedure which incorporates the triazoles via imbibition has been developed (Fletcher and Hofstra, 1990) and used successfully with several crops (Fletcher and Kraus, 1995).

Gibberellins Reverse Triazole-Induced Effects

Inhibition of GA biosynthesis as the primary effect of the triazole PGRs is supported by the evidence that triazole-treated plants have lower concentrations of endogenous GA-like substances (Graebe, 1987; Rademacher, 1997). Furthermore, the PGR, biochemical and physiological properties of triazoles can be reversed by the application of GA. This reversal is independent of the time of application, since similar results were obtained when GA was applied before (Davis, Steffens, and Sankhla, 1988), simultaneously (Gilley and Fletcher, 1998; Fletcher et al., 1988) or after (Davis, Steffens, and Sankhla, 1988) triazole treatment. Based on the interactions of triazoles and GA, it is logical to conclude that the PGR and stress protective effects of the triazoles are a consequence of their primary action as inhibitors of GA biosynthesis.

NON-CHEMICAL REGULATION OF GIBBERELLIN

This chapter has so far provided evidence that GA levels can be reduced using chemical PGRs and this reduction triggers a cascade of events that ultimately leads to stress protection in plants. Secondary effects of PGRs lead to shifts in several other metabolites in the isoprenoid pathway (see Figure 3.1). Despite the "multiprotectant" nature of PGRs, there is a public demand for a reduction of chemicals in the environment, which necessitates the development of alternative technologies for crop protection. Stress tolerance in plants has been improved traditionally through selection of stress-tolerant cultivars, breeding, acclimation (several horticultural plants) and, more recently, by recombinant DNA technology. GAs are ubiquitous in plants and we propose that the primary action of triazole-induced stress protection is mediated by a reduction in GA levels (Fletcher et al., 1999). We have tested the hypothesis that GA-deficient mutants have the ability to tolerate abiotic stresses, using barley (*Hordeum vulgare* L.) as a model system.

A dwarf barley responsive to exogenous GA_3 was produced by gamma irradiation of the cultivar Perth. This dwarf was observed to be tolerant to high temperature and drought stress, compared to its normal near-isogenic counterpart. The application of GA_3 reversed the phenotype and the inherent stress tolerance of the dwarf, whereas the application of PBZ to the normal counterpart induced dwarfism and stress protection. The reversal of the dwarf phenotype by GA_3 application clearly indicates that this dwarf is deficient in GAs. Various GAs were applied to the dwarf seedlings as microdrops and the growth response indicated that the biosynthetic lesion is in the conversion of GA_{20} to GA_1 and GA_9 to GA_4. This study demonstrated for the first time, that modulation of GAs is an attractive approach for conferring stress protection in plants (Vettakkorumakankav et al., 1999). It has been hypothesized (Graebe, 1987) that in most species, the GA responsible for elongation is GA_1, whereas in some species GA_4 plays a role (Hedden and Kamiya, 1997). It is proposed that, in addition to their role in elongation, GA_1 and GA_4 are important for determining stress tolerance in plants. Our observations provide evidence that there is an intimate relationship between GA and stress tolerance. In addition to the use of chemicals, alternate strategies for non-chemical crop protection are illustrated in Figure 3.2. These include classical genetics and breeding strategies using GA mutants with desirable stress tolerance and agronomic properties. GA-deficient mutants from several species have been studied and are available (Phinney, 1984) for use in such breeding programs.

An alternate approach is the transgenic regulation of genes involved in the GA biosynthetic pathway such as *ent*-kaurene synthase or those catalyzing subsequent steps. The enzymes responsible for the production of GA_1 and GA_4 are 2-oxoglutarate-dependent dioxygenases, for which the genes have been cloned from several species (Hedden and Kamiya, 1997) and are attractive sites for gene disruption or antisense regulation. Alternatively, overexpression of the 2β-hydroxylases would result in more GA_1 and GA_4 being inactivated, thereby reducing their levels. Another transgenic strategy is the use of the photoreceptor, phytochrome A (Phy A). The overexpression of oat Phy A in tomato and tobacco has been achieved and these plants resemble GA-deficient plants. Estimation of endogenous levels of GAs in transgenic tobacco overexpressing Phy A indicates that the levels of GA_1 and GA_4 are significantly reduced (Jordan et al., 1995). We have observed that these tobacco plants are more tolerant to free radicals generated by either paraquat or photoinhibition at low temperatures.

SUMMARY

The plant hormone GA is generated by the isoprenoid pathway, which also produces other hormones and metabolites. GA regulates the growth and development of plants from their germination through maturation and senescence. Early PGRs including CCC interfere with *ent*-kaurene synthase and the triazoles with Cyt-P-450-mediated monooxygenase reactions. This interference leads to a reduction in GA and an increase in cytokinin and ABA levels. These hormonal changes lead to several morphological and biochemical effects. Treated plants are shorter with a higher root:shoot ratio and increased levels of photosynthetic pigments. We have demonstrated that triazoles protect plants against various stresses including drought, low and high temperatures, air pollutants and UV-B radiation, and hence have been referred to as "plant multiprotectants." A novel seed treatment procedure that incorporates the triazole by imbibition has been developed in order to minimize the spread of chemicals in the environment. Using this procedure, field experiments indicate that the triazole-induced protection of plants from environmental stresses is maximal under hostile environmental conditions. The cyclohexadiones are a newer generation of GA inhibitors which interfere with the dioxygenases and have PGR and stress protective properties, but are not as effective as the triazoles. Non-chemical procedures for regulating GA levels are transgenics, phytochrome A overexpression, classical genetics and breeding strategies. Our recent studies indicate that the phenotypic and stress tolerant characteristics of a dwarf barley mutant were reversed by GA, whereas the application of a triazole (PBZ) to its normal near-isogenic counterpart induced dwarfism and stress tolerance. Furthermore, GA_1 and GA_4 were more effective than their precursors GA_{20} and GA_9 in reversing dwarfism. From these studies we conclude that modulation of specific GAs, by either chemical or non- chemical procedures is crucial for induction of stress tolerance in plants. However, integration of the various concepts discussed in this chapter are essential and it is proposed that triazoles are a short-term, whereas a breeding program is a long- term plan for crop protection.

REFERENCES

Asare-Boamah, N.K., G. Hofstra, R.A. Fletcher, and E.B. Dumbroff (1986). Triadimefon protects bean plants from water stress through its effects on abscisic acid. *Plant and Cell Physiology* 27:383-390.

Booker, H.M., T.J. Gillespie, G. Hofstra, and R.A. Fletcher (1991). Uniconazole-induced thermotolerance in wheat seedlings is mediated by transpirational cooling. *Physiologia Plantarum* 81:335-342.

Buchenauer, H. and E. Rohner (1981). Effect of triadimefon and triadimenol on growth of various plant species as well as on gibberellin content and sterol metabolism in shoots of barley seedlings. *Pesticide Biochemistry and Physiology* 15:58-70.

Burrows, G.E., T.S. Boag, and W.P. Stewart (1992). Changes in leaf, stem, and root anatomy of Chrysanthemum cv. Lillian Hoek following paclobutrazol application. *Journal of Plant Growth Regulation* 11:189-194.

Cathey, H.M. (1964). Physiology of growth-retarding chemicals. *Annual Review of Plant Physiology* 15:271-284.

Davis, T.D. and E.A. Curry (1991). Chemical regulation of vegetative growth. *Critical Reviews in Plant Science* 10:151-188.

Davis, T.D., G.L. Steffens, and N. Sankhla (1988). Triazole plant growth regulators. *Horticultural Reviews* 10:63-105.

Fletcher R.A. and V. Arnold (1986). Stimulation of cytokinins and chlorophyll synthesis in cucumber cotyledons by triadimefon. *Physiologia Plantarum* 66:197-201.

Fletcher, R.A., N.K. Asare-Boamah, L.C. Krieg, G. Hofstra, and E.B. Dumbroff (1988). Triadimefon stimulates rooting in bean hypocotyl. *Physiologia Plantarum* 73:401-405.

Fletcher, R.A., A. Gilley, T.D. Davis, and N. Sankhla (2000). Triazoles as plant growth regulators and stress protectants. *Horticultural Reviews* 24:55-138.

Fletcher, R.A. and G. Hofstra (1985). Triadimefon—A plant multi-protectant. *Plant and Cell Physiology* 26:775-780.

Fletcher, R.A. and G. Hofstra (1988). Triazoles as potential plant protectants. In *Sterol Synthesis Inhibitors in Plant Protection*, D. Berg and M. Plempel (Eds.). Cambridge: Ellis Horwood Limited, pp. 321-331.

Fletcher, R.A. and G. Hofstra (1990). Improvement of uniconazole-induced protection in wheat seedlings. *Journal of Plant Growth Regulation* 9:207-212.

Fletcher, R.A., G. Hofstra, and J.-G. Gao (1986). Comparative fungitoxic and plant growth regulating properties of triazole derivatives. *Plant and Cell Physiology* 27:367-371.

Fletcher, R.A. and T.E. Kraus (1995). Triazoles: Protecting plants from environmental stress. In *Agri-food Research*, R. Meerveld (Ed.), 18:15-19. Ontario: Queens Printer.

Fletcher, R.A. and V. Nath (1984). Triadimefon reduces transpiration and increases yield in water-stressed plants. *Physiologia Plantarum* 62:422-426.

Fletcher, R.A. and D.J. Osborne (1965). Regulation of protein and nucleic acid synthesis by gibberellin during leaf senescence. *Nature* 207:1176-1177.

Gao, J.-G., G. Hofstra, and R.A. Fletcher (1988). Anatomical changes induced by triazoles in wheat seedlings. *Canadian Journal of Botany* 66:1178-1185.

Gilley, A. and R.A. Fletcher (1998). Gibberellin antagonizes paclobutrazol-induced stress protection in wheat seedlings. *Journal of Plant Physiology* 153:200-207.

Goodwin, T.W and E.I. Mercer (1990). *Introduction to Plant Biochemistry*, Second Edition. Oxford: Pergamon Press.

Graebe, J.E. (1987). Gibberellin biosynthesis and control. *Annual Review of Plant Physiology* 38:419-465.

Grossmann, K. (1992). Plant growth retardants: Their mode of action and benefit for physiological research. In *Progress in Plant Growth Regulation, Proceedings of the 14th International Conference on Plant Growth Substances,* C. Karssen, L. Van Loon, and D. Vreugdenhill (Eds.). Dordrecht, Netherlands: Kluwer Academic Publishers, pp. 788-797.

Grossmann, K., J. Kwiatkowski, C. Häuser, and F. Siefert (1994). Influence of the triazole growth retardant BAS 111W on phytohormone levels in senescing intact pods of oilseed rape. *Plant Growth Regulation* 14:115-118.

Häuser, C., J. Jung, and K. Grossmann (1992). Changes in abscisic acid levels of heterotrophic cell suspension cultures caused by the plant growth retardant BAS 111W and possible physiological consequences. In *Progress in Plant Growth Regulation, Proceedings of the 14th International Conference on Plant Growth Substances,* C. Karssen, L. Van Loon, and D. Vreugdenhill (Eds.). Dordrecht, Netherlands: Kluwer Academic Publishers, pp. 173-179.

Hedden, P. and Y. Kamiya (1997). Gibberellin biosynthesis: Enzymes, genes and their regulation. *Annual Review of Plant Physiology and Plant Molecular Biology* 48:431-460.

Jordan, E.T., P.M. Hatfield, D. Hondred, M. Talon, J.A.D. Zeevaart, and R.D. Vierstra (1995). Phytochrome A overexpression in transgenic tobacco. *Plant Physiology* 107:797-805.

Kraus, T.E., R.C. Evans, R.A. Fletcher, and K.P. Pauls (1995). Paclobutrazol enhances tolerance to increased levels of UV-B radiation in soybean (*Glycine max*) seedlings. *Canadian Journal of Botany* 73:797-806.

Kraus, T.E. and R.A. Fletcher (1994). Paclobutrazol protects wheat seedlings from heat and paraquat injury. Is detoxification of active oxygen involved? *Plant and Cell Physiology* 35:45-52.

Kraus, T.E., B.D. McKersie, and R.A. Fletcher (1995). Paclobutrazol-induced tolerance of wheat leaves to paraquat may involve increased antioxidant enzyme activity. *Journal of Plant Physiology* 145:570-576.

Kraus, T.E., D.P. Murr, and R.A. Fletcher (1991). Uniconazole inhibits stress-induced ethylene in wheat and soybean seedlings. *Journal of Plant Growth Regulation* 10:229-234.

Kraus, T.E., D.P. Murr, G. Hofstra, and R.A. Fletcher (1992). Modulation of ethylene synthesis in acotyledonous soybean and wheat seedlings. *Journal of Plant Growth Regulation* 11:47-53.

Kraus, T.E., K.P. Pauls, and R.A. Fletcher (1995). Paclobutrazol and hardening-induced thermotolerance of wheat: Are heat shock proteins involved? *Plant and Cell Physiology* 26:59-67.

Mackay, C.E., J.C. Hall, G. Hofstra, and R.A. Fletcher (1990). Uniconazole-induced changes in abscisic acid, total amino acids and proline in *Phaseolis vulgaris*. *Pesticide Biochemistry and Physiology* 37:74-82.

Marshall, J.G., J.B. Scarratt, and E.B. Dumbroff (1991). Induction of drought resistance by abscisic acid and paclobutrazol in jack pine. *Tree Physiology* 8:415-421.

Nickell, L.G. (Ed.) (1982). *Plant Growth Regulators: Agricultural Uses*. New York: Springer-Verlag.

Phinney, B.O. (1984). Gibberellin A1 dwarfism and the control of shoot elongation in higher plants. In *The Biosynthesis and Metabolism of Plant Hormones*, A. Crozier and J.R. Hilman (Eds.). *Society for Experimental Biology Seminar Series* 23. Cambridge: Cambridge University Press, pp.19-41.

Rademacher, W. (1997). Bioregulation in crop plants with inhibitors of gibberellin biosynthesis. *Proceedings of the Plant Growth Regulation Society of America, 24th Annual Meeting*. Atlanta, GA, pp. 27-31.

Senaratna, T., C.E. Mackay, B.D. McKersie, and R.A. Fletcher (1988). Triazole-induced chilling tolerance in tomato and its relationship to antioxidant content. *Journal of Plant Physiology* 133:56-61.

Sopher, C.R., M. Król, N.P.A. Huner, A.E. Moore, and R.A. Fletcher (1999). Chloroplastic changes associated with paclobutrazol-induced stress protection in maize seedlings. *Canadian Journal of Botany* 77:279-290.

Stowe, B.B. And T. Yamaki (1957). The history and the physiological action of gibberellins. *Annual Review of Plant Physiology* 8:181-216.

Upadhyaya, A., T.D. Davis, R.H. Walser, A.B. Galbiath, and N. Sankhla (1989). Uniconazole-induced alleviation of low temperature damage in relation to antioxidant activity. *Horticultural Science* 24:955-957.

Vettakkorumakankav, N.N., D. Falk, P.K. Saxena, and R.A. Fletcher (1999). A crucial role for gibberellins in stress protection of plants. *Plant and Cell Physiology* 40:542-548.

Weaver, R.J. (Ed.) (1972). *Plant Growth Substances in Agriculture*. San Francisco, CA: W.H. Freeman and Company.

Webb, J.A. and R.A. Fletcher (1996). Paclobutrazol protects wheat seedlings from injury due to waterlogging. *Plant Growth Regulation* 18:201-206.

Zeevaart, J.A.D. and R.A. Creelman (1988). Metabolism and physiology of abscisic acid. *Annual Review of Plant Physiology* 39:439-475.

Chapter 4

Plant Growth Retardants in Ornamental Horticulture: A Critical Appraisal

Martin P. N. Gent
Richard J. McAvoy

INTRODUCTION

Growth-retardant chemicals are used primarily to control plant height. In this chapter, we will review this use of growth-retardant chemicals on ornamental plants, summarizing the research conducted in the last fifteen years. We refer to studies of other plants, such as cereal crops and fruit trees, only where they are the subject of more sophisticated physiological investigations. We list the current commercial recommendations for application of the five most commonly used growth-retardant chemicals. Except for daminozide, all these chemicals are gibberellin biosynthesis inhibitors. The triazole type of gibberellin biosynthesis inhibitors, an important new class of chemicals, has been developed since 1980. This type was not included in the last comprehensive review of plant growth-retarding activities of five chemicals on a wide variety of ornamental plant species (Cathey, 1975). A more recent review of triazole growth retardants in horticulture (Davis, Steffens, and Sankhla, 1988) included only the early references. Other reviews of growth retardants have focused on mode of action (Cathey, 1964), benefit for physiological research (Grossmann, 1992), and applications in citriculture (Harty, 1988) and, more generally, in agriculture and horticulture (Gianfagna, 1995; Rademacher, 1991; Sachs and Hackett, 1972).

A related area is gibberellin (GA) biosynthesis. A book describes GA metabolism and the role of GA in governing plant growth (Takahashi, Phinney, and Macmillan, 1991). Other reviews have focused on gibberellins and reproductive development in seed plants (Pharis and King, 1985), and gibberellin biosynthesis enzymes and their regulation (Hedden and Kamiya, 1997). These reviews elaborate the complex sequence of metabolic steps leading to the production of the various closely related gibberellin structures found in plants.

Growth-retardant chemicals are used primarily to control stem elongation of ornamental plants while they are growing in containers rather than after the plants are transplanted into the landscape. In production of horticultural crops, high fertility and an unlimited supply of water are used to enhance plant growth. This often leads to a plant that is taller than desired. Photoperiod manipulation is used to control flower induction in some species. Increased stem elongation can be a consequence when artificial light is used to produce a long photoperiod. In addition, the warm temperatures required to get a crop to flower on schedule can cause plants to stretch, regardless of photoperiod. Growth retardants can be used in each of these situations to achieve an ideal plant height. In contrast to agronomy, where growth retardants are applied to few species, in ornamental horticulture, growth retardants are used on many species with a diversity of growth habits. These include annuals, perennials, bulbs, and woody shrubs grown primarily for their flowers, as well as tropical plants and shrubs grown primarily for the appearance of foliage. In summarizing the work on this diverse group of plants, it is apparent that plants have a wide variety of responses to growth retardants. In some cases, these chemicals induce more leaves, in others less. In some cases, the area of individual leaves increases, while in others it decreases. In some species, flowering is promoted, and in others it is delayed. Part of the difference in response among plant species is due to different growth habits, and part is due to differences in metabolism and response to gibberellin. In addition, the various gibberellin biosynthesis inhibitor chemicals do not affect the concentration of all gibberellins equally. In the future, variations within and among species in response to different growth-retardant chemicals may become an exciting area of new applications in ornamental horticulture.

GROWTH-RETARDANT CHEMICALS
AND THEIR MODE OF ACTION

Five growth-retardant chemicals are commonly used in ornamental horticulture. We focus primarily on these growth retardants. The introduction of chemicals containing the triazole moiety, in particular paclobutrazol and uniconazole, has inspired much recent research. These chemicals are effective at low concentrations and show few side effects compared to chlormequat and daminozide, the two most widely used chemicals. Ancymidol, another chemical widely used as a growth retardant in ornamental horticulture, is also effective at low concentrations. Table 4.1 lists these chemicals, their common names, chemical names, and trade names. The table lists other chemicals mentioned in plant growth-retardant research and indicates whether they are experimental or registered for commercial use. The table also lists some triazole compounds that are labeled for use as fungicides rather than growth retardants because they have growth-retardant properties. This section summarizes the biochemical modes of action of these chemicals.

Daminozide

Daminozide was one of the first chemicals used to inhibit plant growth (Cathey, 1975). Its chemical structure is unique among the growth retardants, and there are no other analogous chemicals in commercial use for this purpose. Its mode of action is either to inhibit translocation of gibberellins or to promote their degradation (Rademacher, 1991). However, a recent study suggests that daminozide and prohexadione may have similar modes of action as inhibitors of the late stages of gibberellin metabolism (Brown, Kawaide, and Yang, 1997). Because of its low cost and long history of use, this chemical is widely used in all aspects of horticulture. In the United States, daminozide accounts for 54 percent by weight of the growth regulator used in greenhouses (Garber et al., 1996). At a given dose, other chemicals far more effectively inhibit stem elongation than daminozide, but their cost-effectiveness has not yet been demonstrated.

TABLE 4.1. Common, Chemical, and Trade Names of Growth-Retardant Chemicals

Common Name	Chemical Name	Trade Names
	Gibberellin biosynthesis inhibitors	
daminozide	butanedioic acid mono-[2,2-dimethyl-hydrazide], succinic acid-[2,2-dimethyl-hydrazide]	Alar, B-Nine, SADH
chlormequat	2-Chloroethyl-N,N,N-trimethylammonium chloride	Cycocel, CCC, Cycogan
mepiquat	1,1-dimethyl piperidinium chloride	Pix
ancymidol	α-cyclopropyl-α-[p-methoxyphenyl]-5-pyrimidinemethanol	A-Rest, EL531, Reducymol
flurprimidol	α-(1-methylethyl)-α-[4-(trifluoromethoxy) phenyl]-5-pyrimidinemethanol	Cutless, EL500
inabenfide	4'-chloro-2'-(α-hydroxybenzyl) isonicotinanilide	Seritard
prohexadione	calcium 3-oxido-5-oxo-4-propionylcyclohex-3-ene carboxylate	Viviful
primo	4-(cyclopropyl-α-hydroxymethylene)-3,5-dioxocyclohexanecarboxylate ethyl ester	CGA163935, experimental
tetcyclacis	5-(4-chlorophenyl)-3,4,6,9,10-pentaaza-tetracyclo-5,4,$10^{2,6}$,$0^{8,11}$ dodeca-4-9-diene	Kenbyo, BAS106, LAB102883
	Triazole-containing compounds	
paclobutrazol	(2RS,3RS)-1-[4-chlorophenyl]-4,4-dimethyl-2-[1H-1,2,4-triazol-1-yl]-pentan-3-ol	Bonzi, Clipper, Cultar, PP333
uniconazole	(E)-1-[4-chlorophenyl]-4,4-dimethyl-2-[1,2,4-triazol-1-yl]-1-penten-3-ol	Sumagic, XE1019, S3307
triadimefon	1-[4-chlorophenyl]-3,3-dimethyl-1-[1H-1,2,4-triazol-1-yl]-butan-2-one	Amiral, Bayleton
propiconazol	1-[[2-(2,4-dichlorophenyl)-4-propyl-1,3-dioxolan-2-yl]-methyl]-1H-1,2,4-triazole	Banner, Tilt
triapenthanol	β-(cyclohexylmethylene)-α-(1,1-dimethylethyl)-1H-1,2,4-triazol-1-ethanol	Baronet, RSW0411
	1-phenoxy-5,5-dimethyl-3-(1,2,4-triazol-1-yl)-hexan-5-ol	BAS111, experimental
	Other growth regulators	
ethephon	2-chloroethyl phosphoric acid	Florel, Madurex
maleic-hydrazide	1,2-dihydropyridazine-3,6-dione, 6-hydroxy-2H-pryidazin-3-one	Retard, De-Cut, Royal Slo-Gro
chlorflurenol	methyl-2-chloro-9-hydroxyflurene-9-carboxylate	Maintain
dikegulac	sodium 2,3:4,6-bis-O-(1-methylethylidene)-α-L-xylo-2-hexalofuranosonic acid	Atrinal, Atrimmec
mefluidide	N-[2,4-dimethyl-5-[[(trifluoromethyl) sulfonyl] amino] phenyl] acetamide	Embark, Trim-Cut
dimethipin	2,3-dihydro-5,6-dimethyl-1,4-dithiin-1,1,4,4-tetraoxide	Harvade
oxathiin	2,3-dihydro-5,6-diphenyl-1,4-dithiin-1,1,4,4-tetraoxide	UPIP293, experimental

Onium Compounds

Chlormequat and mepiquat are two other chemicals in wide use. These compounds contain a quaternary ammonium group. They inhibit the step in the biosynthesis of gibberellins from geranylgeranyl pyrophosphate to *ent*-kaurene (Rademacher, 1991). They may also inhibit sterol synthesis, as *ent*-kaurene is also the precursor in this biochemical pathway. These growth retardants are specific for one of three *ent*-kaurene synthases in plants (Barrett, 1982). Chlormequat is used on a variety of ornamental crops, and in terms of amount of chemical used, it is the second most common growth retardant used by the greenhouse industry in the United States (Garber et al., 1996). The major use of mepiquat is as a growth retardant in cotton (Reddy, Reddy, and Hodges, 1996; Zhang, Cothren, and Lorenz, 1990).

Pyrimidine Analogs

The chemicals ancymidol and flurprimidol contain a pyrimidine group and are potent inhibitors of gibberellin biosynthesis. A dose of ancymidol need be only 1/100 to 1/1000 of that of chlormequat to get a similar growth response. These compounds block the action of a monooxygenase enzyme that catalyzes *ent*-kaurene oxidation, a necessary step in the pathway between *ent*-kaurene and gibberellins (Rademacher, 1991). Ancymidol is used primarily to retard stem elongation of annuals and perennials grown in containers (Tija, 1976; Whipker, Eddy, and Hammer, 1995; Whipker et al., 1995; Cramer and Bridgen, 1998). Flurprimidol is primarily used to regulate growth of trees in the landscape (Gilliam, Fare, and Eason, 1988; Sterrett, Tworkoski, and Kujawski, 1989; Norcini, 1991; Arron et al., 1997). It may have application for long-term inhibition of ornamental woody shrubs (Keever, Gilliam, and Eakes, 1994).

Other Compounds

Inabenfide, a nicotinic acid derivative, and tetcyclacis, an experimental norbornadiazo compound, are heterocyclic compounds with multiple nitrogen substitution in their rings. These also block *ent*-kaurene oxidation by the mechanism described for pyrimidine-containing compounds (Rademacher, 1992b). Their efficacy has been demonstrated on various ornamental plants (Holcomb, Ream, and Reed, 1983; McDaniel, 1986). The most recently developed class of growth retardants are the cyclohexanetriones, which include the compound prohexadione. These appear to block a late step in gibberellin metabolism, in particular, the step from GA_{20} to GA_1 (Brown, Kawaide, and Yang, 1997). These compounds appear to act as a 2-oxoglutarate substitute and to inhibit anthocyanin formation (Rademacher, 1992a).

Triazole Compounds

The triazole chemicals are related to fungicides that act primarily as inhibitors of one or more cytochrome oxidase enzymes that are important in biosynthesis of sterols (Rademacher, 1991; Benton and Cobb, 1997), a necessary component of fungal cell walls. When applied as a fungicide to cereal crops, triadimefon had a side benefit of reducing stem elongation. A search for similar chemicals with more potent growth regulator activity resulted in the development of paclobutrazol and uniconazole. Inspired by their high efficacy, numerous reports of the response of ornamental plants to paclobutrazol and uniconazole have been made in recent years (Davis, Steffens and Sankhla, 1988; Barrett and Nell, 1989; Cole and Frymire, 1995). The triazoles have both antigibberellin and antisterol synthesis properties, and the specificity depends on minor modification in the chemical structure (Rademacher, 1991). For instance, paclobutrazol and uniconazole can exist in different enantiomers or stereoisomers with the same chemical structure, and these enantiomers differ in their biological activity as sterol biosynthesis inhibitors (Sugavanam, 1984; Izumi et al., 1985; Haughan et al., 1989; Lenton, Appleford, and Templesmith, 1994). Perhaps one component of the efficacy of paclobutrazol and uniconazole as plant growth retardants is that they block multiple steps in the biosynthesis of gibberellins and sterols. However, it is still an open question how much these two activities play a part in suppression of plant growth. For instance, pea (*Pisum sativum* L.) showed inconsistency between changes in growth and endogenous levels of gibberellins and sterols when plants were treated with the different uniconazole enantiomers (Yokota et al., 1991).

Reversal with Gibberellic Acid

Synthetic gibberellins, namely GA_3 and GA_{4+7}, are used commercially to control plant growth and development. Application of these synthetic gibberellins often results in a plant response identical to that due to endogenous gibberellins (Takahashi, Phinney, and Macmillan, 1991), and they can reverse some of the effects of gibberellin biosynthesis inhibitors. GA_3 reversed the effect of accelerated flowering of *Pelargonium* ×*hortorum* L.H. Bailey 'Sprinter Scarlet' caused by chlormequat (Armitage, 1986). Gibberellins reversed the effect of excess paclobutrazol on stem elongation of *Pelargonium* 'Mustang' if applied one day after the growth retardant, but not if applied fourteen to twenty-one days later (Cox, 1991). Gibberellic acid reversed the effect of uniconazole on stem elongation in *Dendranthema* ×*grandiflorum* Ramat 'Echo' if applied within three weeks, but not if applied four to five weeks after uniconazole (Holcomb, Tukey, and Rose, 1991). A 0.2 milligram (mg) drench of uniconazole followed by one 100 parts per million (ppm), or 4×25 ppm, spray of GA_{4+7} partly restored stem elongation and stimulated flowering of *Hibiscus rosa-sinensis* L. 'Jane Cowl' (Wang and Dunlap, 1994).

Pinching Agents

Some growth-retardant chemicals prevent growth of axillary branches (suckers). These include maleichydrazide and a class of chemicals called oxathiins. The chemicals dikegulac and chlorflurenol are potent inhibitors of tree growth, and also fall into this class of chemical pinching agents. The primary mode of action of these chemicals is to prevent cell development, disrupt differentiation in the meristem, and repress apical dominance (Arzee, Langenauer, and Gressel, 1977). Trunk injection of dikegulac can reduce regrowth of deciduous trees for two or three years (Domir and Roberts, 1981, 1983). A 5,000 ppm spray of dikegulac was more effective at controlling growth of large established shrubs than a 500 ppm spray of flurprimidol (Banco and Stefani, 1996). A dikegulac spray on *Murraya paniculata* Jack. decreased apical dominance among lateral shoots (Kawabata and Criley, 1996) and increased the number of lateral shoots on *Rhododendron calendulaceum* Michx. (Malek et al., 1992). Oxathiin inhibited stem elongation and promoted lateral branches of *Dendranthema* ×*grandiflorum* 'Paragon' (Purohit and Shanks, 1984). However, 250 to 1,000 ppm dikegulac did not increase basal shoots of most cultivars of *Rosa odorata* L. (Jayroe-Counoyer and Newman, 1995). A 1,600 ppm dikegulac spray increased flowering and reduced size only in one of five cultivars of *Bougainvillea glabra* Choicy (Norcini, Aldrich, and McDowell, 1994). Surfactants affect the efficacy of maleic hydrazide, especially on *Pinus*, and also on *Xylosma congestum* Lour., *Viburnum japonicum* Spreng., *Juniperus*, *Pyracantha coccinea* Roem., but not on *Cotoneaster pannosa* Franch. or *Nerium oleander* L. (Sachs et al., 1975). These chemicals can damage the foliage, so they are not in wide use as growth retardants for ornamental plants. Dikegulac injection induced wilting, curling, and marginal necrosis in *Platanus aceribolia* L., and in *Acer rubrum* and *Acer platanoides* trees (Wright and Moran, 1988). Trunk bark banding with chlorflurenol applied late in the season to deciduous trees deformed leaf blades the following spring (Hield, 1979). A spray of 0.1 to 1.0 percent dikegulac inhibited shoot growth for three months in *Xylosma, Pyracantha, Callistemon citrinus* Curt., or *Cotoneaster,* but it caused chlorosis of *Nerium* unless plants were pruned before spraying (Sachs et al., 1975).

Ethephon is used in the ornamental industry to delay flowering, selectively abort flowers, abscise leaves, as well as to reduce stem elongation and increase stem strength. Within plant tissue, ethephon decomposes into the hormone ethylene plus chloride and phosphate ions. The commercial product Florel is labeled for use on ornamentals at 500 to 1,000 ppm to increase axillary branching on *Pelargonium* ×*hortorum, P. peltatum, Rhododendron* (azalea), *Lantana* spp., *Fuchsia* ×*hybrida, Dendranthema*, and *Vinca major* L. At the proper concentration, flower buds are selectively aborted, but apical buds and leaves are not. Ethephon can suppress flowering in plants used for vegetative propagation, or delay flowering under naturally inductive conditions. A single concentrated drop, 10 microliters (µl) of 10,000 ppm ethephon, applied to leaves of *Dendranthema* increased leaf number, delayed flowering under inductive conditions, reduced stem elongation, reduced apical dominance, and increased axillary shoot growth (Stanley and Cockshull, 1989), even when the leaf was excised 12 hours (h) after the treatment was applied. This suggests ethylene was produced endogenously even after the leaf was removed. The response to ethephon was greater for the cultivar Polaris than for 'Bright

Golden Anne'. All flowers on ethephon-treated racemes of *Strongylo-don macrobotrys* abscised within 75 h of treatment, while flowers on untreated racemes aborted after 190 h (Furutani, Johnston, and Nagao, 1989). The sensitivity to abscission due to ethephon varied among plant organs (Woolf, Clemens, and Plummer, 1992). Floral buds of *Camellia sasanqua* Thunb. were more sensitive than leaves and vegetative buds to a 1,000 ppm spray of ethephon; this treatment aborted 40 to 80 percent of flower buds, but only 5 to 10 percent of vegetative buds. Leaves on newly extended shoots were more sensitive to ethephon than fully expanded leaves, and sensitivity of all *Camellia* plant parts to ethephon increased with temperature (Woolf, Clemens, and Plummer, 1995).

Ethephon is used to control height and increase stem strength in *Narcissus* and *Hyacinthus*. Ethephon controlled stem length without affecting flower development in *Pelargonium* ×*hortorum* 'Yours Truly' (Tayama and Carver, 1990). A 500 ppm spray of ethephon applied to trunks of *Prunus dulcis* L. increased cambial strength and reduced bark injury typically associated with mechanical shaking (Gurusinghe and Shackel, 1995), coincident with an increase in area and thickening of tangential ray initial cell walls. Ethephon increased phenylalanine-ammonia-lyase activity and subsequent anthocyanin content in the skin of fruit of *Malus domestica* Borkh L., resulting in color improvement during both the preharvest and cold-storage periods (Larrigaudiere, Pinto, and Vendrell, 1996).

CONTROL OF PLANT HEIGHT

The primary use of growth-retardant chemicals is to control the height of plants. As a plant hormone, gibberellin induces the elongation of individual plant cells. Application of GA_1 or GA_{4+7} results in tall plants with long internodes (Takahashi, Phinney, and Macmillan, 1991). The gibberellin biosynthesis inhibitors lower the concentration of gibberellins in plants and inhibit cell expansion by lowering the gibberellin-dependent cell wall relaxation (Cosgrove and Sovonick-Dunford, 1989). Thus, the primary effect of growth retardant is to inhibit internode elongation. This is observed in all types of plants: annuals, seedlings, herbaceous perennials, foliage plants, and established trees. The only reported exception is *Hedera helix* L. (Horrell,

Jameson, and Bannister, 1990). Ancymidol, tetcyclacis, and paclobutrazol can also affect phototropism of *Vigna radiata* L., perhaps due to their effect on a cytochrome P-450 oxygenase enzyme, while chlormequat is not effective (Konjevic, Grubisic, and Neskovic, 1989). The site of application can have a dramatic effect on plant response. Paclobutrazol applied before floral induction to the base of new shoots of *Pyrus communis* L. stimulated flower number without inhibiting shoot growth (Browning, Kuden, and Blake, 1992), while 30 ppm applied to shoot apices inhibited growth as much as a 1,000 ppm whole-plant spray (Browning et al., 1992).

General Recommendations

The ornamental plants can be separated into several classes: bedding plants, annual flowers, perennial flowers, bulbs, woody shrubs, and foliage plants. Labels for the commercially available growth retardants, A-Rest (ancymidol), Cycocel (chlormequat), B-Nine (daminozide), and Bonzi (paclobutrazol), allow users to experiment with these chemicals within these classes of plants, even if the species is not specifically identified on the label. These general label recommendations for spray or drench applications are summarized in Table 4.2, with the recommendations given as concentrations. Table 4.2 illustrates the wide range in efficacy of the commercially available growth-retardant chemicals. Ancymidol and paclobutrazol have similar efficacy; a spray of less than 50 ppm or a drench of about 1 ppm provides a 25 to 33 percent reduction in stem elongation. However, to provide the same effect, spray concentrations need to be about 40 times greater for chlormequat, and 100 times greater for daminozide. Although not listed in Table 4.2, uniconazole efficacy is about fourfold that of paclobutrazol on *Dendranthema* 'Bright Golden Anne' (Gilbertz, 1992). Uniconazole at 10 ppm was equivalent to chlormequat at 1,500 ppm on *Pelargonium* 'Yours Truly', and 20 ppm on *Dendranthema* was equivalent to 5,000 ppm daminozide (Tayama and Carver, 1992a, b), or a 0.45 mg ancymidol drench (Starman, 1990).

An implicit assumption in the recommendations for spray application is that larger plants will receive a greater volume of spray than smaller plants because they have a larger leaf surface area and occupy a larger area of bench space. The recommended volumes of spray per unit bench or ground area vary among the products; 100 milliliters per

TABLE 4.2. Test Rates to Be Used on Species Not Specifically Listed on Product Labels, but Where Use Is Permitted on a General Group of Ornamentals

Ornamental Crop	Ancymidol (A-Rest)	Chlormequat (Cycocel)	Daminozide (B-Nine)	Paclobutrazol (Bonzi)
	Concentrations (mg·liter^{-1} of active ingredient) and method of application*			
Bedding plants in plugs	3-35S, 0.5-1D	400-750S	1,500-2,500S	1-20S
Bedding plants in final containers	6-66S, 1-2D	800-1,500S	2,500-5,000S	5-90S, 0.5-1D
Bulb crops	25-50S, 2D			50S, 5D, 10BS
Foliage plants	20-50S, 1-2D		2,500-7,500S	15-30S, 0.5-1D
Herbaceous perennials	20-50S, 1-2D	1,250S, 2,000-4,000D		15-30S, 0.5-1D
Flowering woody shrubs	50S, 2D	800-4,000S, 2,000-4,000D		50S, 2D
Landscape woody shrubs		800-4,000S, 2,000-4,000D		100S, 4D

Note: Drench treatments are typically applied at 65 ml·liter^{-1} of potting medium. Recommended carrier volumes for sprays are 100 ml·m^{-2} for ancymidol and daminozide and 200 to 300 ml·m^{-2} for chlormequat and paclobutrazol.

* Application methods include foliar spray (**S**), soil drench (**D**), or bulb soak (**BS**).

square meter (ml·m^{-2}) for ancymidol and daminozide, 200 ml·m^{-2} for uniconazole, and 200 to 300 ml·m^{-2} for chlormequat and paclobutrazol. The product label recommendation for a root medium drench application is also given as a concentration. The product labels assume a standard delivery volume for drench application that allows the solution to be uniformly distributed throughout the entire volume of a moist potting medium. The recommendation for all products is approximately 65 milliliters per liter (ml·liter^{-1}) of potting medium. These conversion factors can be used to determine the dose corresponding to a given concentration. The dose is the concentration of active ingredient times the solution volume applied per plant. In experimental research reports, drench applications are often a dose of active ingredient per plant or per pot because plant response to a growth-retardant chemical is related to the dose received in milligrams of active ingredient per plant (Sanderson, Martin, and McGuire, 1988). For instance, the response to uniconazole of *Dendranthema* 'Ovaro' and various *Euphorbia pulcherrima* Willd. cultivars did not vary with carrier volume as long as dosage remained constant (Bailey, 1989a, b). The use of pot covers demonstrated that any increased effect of uniconazole spray observed with increased carrier volume was due to growth-retardant solution that dripped into the rooting medium (Barrett, Bartuska, and Nell, 1994a, b). The pot covers negated these effects.

Recommendations by Species

The commercial products provide specific recommendations for a large number of ornamental plant species. These are summarized in Table 4.3, along with experimental results for species not listed on the labels. In addition to spray and drench applications, concentrations are recommended for a preplant dip of cuttings or bulbs. Even for a particular species, the label recommendations provide a range of rates. This is because more rapidly growing plants require higher doses of growth retardants to achieve desirable control, while plants growing under suboptimal conditions require lower ones. Even when paclobutrazol was applied to well watered and fertilized *Dendranthema* 'Nob Hill' and *Petunia hybrida* 'White Flash', the plants were larger than those allowed to wilt or those receiving less fertilizer (Barrett and Nell, 1990). Similar effects were observed with well-fertilized *Dieffenbachia maculata* L. treated with ancymidol (Joiner et al., 1978). The amount of ancymidol required to reduce internode

TABLE 4.3. Recommended Application Rates for Various Ornamental Species from Commercial Product Labels or Published Data

Ornamental Crop	Ancymidol (A-Rest)	Chlormequat (Cycocel)	Daminozide (B-Nine)	Paclobutrazol (Bonzi)	Uniconazole (Sumagic)	Experimental Reference
		Concentrations (mg·liter^{-1} of active ingredient) and method of application*				
Achimenes hybrids		1,250S, 2,000-4,000D				
Ageratum houstonianum	10-15S+	800-1,500S	2,500-5,000S	15-45S	20-30S	
Alcea rosea				50SX, 2DX		Latimer, Oetting, and Thomas (1995)
Altemanthera bettzickiana	25-13S, 2-4D					
Antirrhinum majus	15-20S+			45-90S	25-50S	
Aquilegia x hybrida and Aquilegia spp.	65-13S, 2-4D	1,250S, 2,000-4,000D				
Aster novi-belgii and Aster spp.		1,250S, 2,000-4,000D	5,000SX	160SX	20SX	Whipker et al. (1995)
Balena cristata		1,250S, 2,000-4,000D				
Begonia x semperflorens-cultorum		800-4,000S, 2,000-4,000D				
Begonia x semperflorens-cultorum	6-12S+		2,500-5,000S	**	**	
Begonia x tuberhybrida, B. x hiemalis, B. rex		800-1,500S				
Bougainvillea glabra		1,250S, 2,000-4,000D		50S, 2D		Wilkinson and Richards (1987)
Bouvardia humboldtii				1,000SX, 25DX		
Brassaia actinophylla				3-8DX	0.4-0.7DX	Barrett, Bartuska, and Nell (1994b); Wang and Blessington (1990)
Brassica oleracea			2,500-5,000SX	15SX	5SX	Whipker et al. (1995)
Browallia speciosa			2,500-5,000S			
Caladium x hortulanum				100-200S, 2-16D, 60BS		
Calendula officinalis			2,500-5,000S			
Callistephus chinensis	10-15S+		2,500-5,000S			
Camellia x williamsii		1,250S, 2,000-4,000D		50S, 2D		

101

TABLE 4.3 (continued)

Ornamental Crop	Ancymidol (A-Rest)	Chlormequat (Cycocel)	Daminozide (B-Nine)	Paclobutrazol (Bonzi)	Uniconazole (Sumagic)	Experimental Reference
	Concentrations (mg·liter⁻¹ of active ingredient) and method of application*					
Capsicum annuum					25-50DX	Starman (1993)
Catharanthus roseus and hybrids	8-13S+	800-1,500S	2,500-5,000S		1-3S	
Celosia argentea	10-15S+	800-1,500S	2,500S	15-45S	10-20S	
Centurea cyanus and Centurea spp.	10-15S+		2,500-5,000S			
Clematis tangutica	25-132S, 2-4D					
Cleome spinosa	10-15S+	800-1,500S				
Clerodendron thomasoniae	5DX			4DX		Sanderson, Martin, and Reed (1989); Sanderson, Martin, and McGuire (1990)
Codiaeum variegatum				12SX, 1-4DX	0.7-2.5DX	Barrett, Bartuska, and Nell (1994b); Wang and Blessington (1990)
Coleus blumei	10-15S+	800-1,500S	2,500-5,000S	15-45S	10-20S	
Convolvulus spp.		1,250S, 2,000-4,000D				
Cordyline terminalis				200SX, 200DX		Hagiladi and Watad (1992)
Cosmos bipinnatus			2,500-5,000S			
Crossandra infundibuliformis			2,500-5,000S			
Dahlia pinnata (seed)	10-15S+	800-1,500S	2,500-5,000S	15-45S	10-20S	
Dahlia variabilis (rhizome)	2-4D			10-40D		
Delphinium spp.	35-132S, 2-4D		2,500-5,000S			
Dendranthema x grandiflora	25-50S, 2-4D	1,250S, 2,000-4,000D	1,250-5,000S, 1,000DP	50-200S, 14D	2.5-10S	
Dianthus caryophyllus		1,250S, 2,000-4,000D				
Dianthus spp.	10-15S+	800-1,500S	2,500-5,000S	20-60S		
Dicentra spectabilis	65-132S, 2-4D					
Diefenbachia maculata	1-4DX					
Dracaena deremensis var Warneckii and D. spp.	25-132S, 2-4D					Joiner et al. (1978)
Epipremnum aureum	25-132S, 2-4D		2,500-7,500S	50-200SX, 1-1.5DX	25SX, 0.5-1DX	Wang, Hsiao, and Gregg (1992); Wang and Gregg (1994)
Episcia cupreata				7DX		Stamps and Henny (1986)
Euonymus spp.				100S, 4D		

Ornamental Crop	Ancymidol (A-Rest)	Chlormequat (Cycocel)	Daminozide (B-Nine)	Paclobutrazol (Bonzi)	Uniconazole (Sumagic)	Experimental Reference
	Concentrations (mg·liter⁻¹ of active ingredient) and method of application*					
Euphorbia pulcherrima			2,500-5,000S			
Eustoma grandiflorum	65-132S, 2-4D					
Exacum			2,500-5,000S			
Fatshedera lizei	65-132S, 2-4D					
Ficus spp.			2,500-7,500S	0.1-0.2DX		LeCain, Schekel, and Wample (1986)
Forsythia spp.					50SX, 0.5DX	Warren (1990)
Freesia hybrida	3DX			100-300BS		Wulster and Gianfagna (1991)
Fuchsia x hybrida		1,250S, 2,000-4,000D		2.5DX		Roberts, Eaton, and Seywerd (1990)
Gardenia jasminoides	50S, 2D	1,250S, 2,000-4,000D	5,000S			
Gerbera jamesonii	25-132S, 2-4D		1,200-5,000S			
Gladiolus x hortulanus	6-12DX					Shaw, Schekel, and Lohr (1991)
Gomphrena globosa		800-1,500S	2,500-5,000S			
Gynura aurantiaca	25-132S, 2-4D	1,250S, 2,000-4,000D				
Hedera helix		1,250S, 2,000-4,000D				
Helianthus annuus	66SX	800-1,500S		15-60DX		Shravan, Evans, and Whipker (1998); Starman, Kelly, and Pemberton (1989)
Hibiscus rosa-sinensis		200-600S	2,500-5,000S	30-150S, 1-2D	5-15SX, 0.4-1.6DX	Newman, Tenney, and Follett (1989); Wang and Gregg (1989)
Hippeastrum vittatum hybrids				200D		
Hydrangea macrophylla		1,250S, 2,000-4,000D	5,000S		10SX	Bailey (1989a)
Hypoestes phyllostachya		800-1,500S			5-10SX, 0.4-0.8DX	Starman and Gibson (1992)
Ilex spp.	50S, 2D	800-4,000S, 2,000-4,000D		100S, 4D	25SX, 2-4DX	Frymire and Cole (1992); Norcini and Knox (1989a)
Impatiens wallerana	20-26S+			15-45S	5-10S	
Ixora spp.				100S, 4D	15S	
Juniperus spp.						
Kalanchoe blossfeldiana		1,250S, 2,000-4,000D	1,200-5,000S	10-30SX	4-12SX	Gent (1995)
Kalmia spp.					50S, 2DX	Warren (1990)
Lagerstroemia indica					100-200SX, 4-8DX	Ruter (1996)
Lantana camara	25-132S, 2-4D					

TABLE 4.3 *(continued)*

Ornamental Crop	Ancymidol (A-Rest)	Chlormequat (Cycocel)	Daminozide (B-Nine)	Paclobutrazol (Bonzi)	Uniconazole (Sumagic)	Experimental Reference
	Concentrations (mg·liter^{-1} of active ingredient) and method of application*					
Liatris spicata	25-132S, 2-4D					
Ligustrum spp.				100S, 4D	25SX	Nordini and Knox (1989a)
Lilium longiflorum	30-132S, 2-4D	1,250S, 2,000-4,000D			10-25S, 0.25-0.5D	
Lilium spp. (hybrid lilies)	1-2DX	1,250S, 2,000-4,000D		250-500S, 4-30D, 20-30BS	10-50SX, 1-5DX	Bailey and Miller (1989a); Jiao, Wang, and Tsujita (1990)
Magnolia sp.				100S, 4D		
Mandevilla sp.					5-20SX	Deneke, Keever, and McGuire (1992)
Matthiola incana				10DX		Ecker et al. (1992)
Monstera deliciosa	25-132S, 2-4D					
Narcissus	25-132S, 2-4D			20-40D		
Nephthytis gravenreuthii		1,250S, 2,000-4,000D				
Pachystachys coccinea					25-50SX	Hamada, Hosoki and Maeda (1990)
Paeonia sulfruticosa	33-66S	1,500S		10-30S	2-4S	
Pelargonium x hortorum (seed)	33-66S			10-30S	3-6S	
P. x hortorum (vegetative)	15-20S+	800-1,500S				
Pentas lanceolata	25-132S, 2-4D	1,250S, 2,000-4,000D				
Petunia x hybrida			2,500-5,000S		25S, 2-4DX	Frymire and Cole (1992)
Philodendron scandens			2,500-7,500S	30-60S	25-50S	
Phlox drummondii			2,500-5,000S			
Photinia spp.				100S, 4D	45-75DX	Deneke, Thomas, and Keever (1992)
Physostegia virginiana	25-132S, 2-4D	1,250S, 2,000-4,000D		25-50SX	25SX	Cox and Wittington (1988)
Pilea cadierei	100SX			100S, 4D		Scnurr, Cheng, and Boe (1996)
Pinus spp.				1.3DX	0.1DX	Wang and Blessington (1990)
Plectranthus australis	10-15S+					
Portulaca grandiflora		800-4,000S, 2,000-4,000D				
Pseuderanthemum lactifolia					25S, 1DX	Henderson and Nichols (1991); Warren (1990)
Pyracantha spp.						
Rhododendron spp. and hybrids and R. obtusum	26S	1,000-2,000S	2,500S	100-200S, 5-15D	10-15S	

Ornamental Crop	Ancymidol (A-Rest)	Chlormequat (Cycocel)	Daminozide (B-Nine)	Paclobutrazol (Bonzi)	Uniconazole (Sumagic)	Experimental Reference
	Concentrations (mg·liter⁻¹ of active ingredient) and method of application*					
Rosa hybrids (potted)		1,250S, 2,000-4,000D				
Rudbeckia hirta	50DX					Orvos and Lyons (1989)
Salpiglossis sinuata			1,000-5,000SX			Needham and Hammer (1990)
Salvia splendens and Salvia spp.	18-26S+	1,250S, 2,000-4,000D	2,500-5,000S	20-60S	5-10S	
Schefflera arboricola	25-132S, 2-4D	1,250S, 2,000-4,000D				
Sedum spp.			2,500-5,000S			
Senecio cineraria		1,250S, 2,000-4,000D				
Senecio cruentus						
Sinningia speciosa			1,250S			
Solanum pseudo-capsicum		800-1,500S				
Syngonium podophyllum	6-18DX		2,500-7,500S	1.4-2.8DX	0.7DX	Wang (1987); Wang and Blessington (1990)
Tagetes erecta and Tagetes patula	18-26S+	800-1,500S	2,500-5,000S	15-60S, 10-20DX	10-20S	Keever and Cox (1989)
Tibouchina urvilleana		3,000-6,000DX		2.5DX		Roberts, Eaton, and Seywerd (1990)
Tritonia crocata (Ixia)				20-30BS		
Tropaeolum majus		800-1,500S				
Tulipa gesneriana	1-4D			5-40D, 2-5BS		
Verbena x hybrida		800-1,500S	2,500-5,000S			
Viola spp.					1-5S	
Viola x wittrockiana	8-10S+			5-15S	1-6S***	
Zantedeschia spp.				10-30D, 20BS		
Zinnia elegans	10-15S+	800-1,500S	2,500-5,000S	15-45S		

* Application methods include foliar spray (S), soil drench (D), bulb soak (BS), or cutting dip (DP). X denotes experimental rates used to achieve a 25 to 50 percent reduction in elongation. + denotes the use of a 50 percent higher concentration to hold finished plants.

** Denotes extreme sensitivity to a growth regulator; do not apply to this crop and avoid spray drift.

*** Denotes high sensitivity to a growth regulator; do not apply to plants less than 10 cm tall. Recommended volumes are for drench applications, 65 ml·liter⁻¹ of potting medium; for spray applications, 100 ml·m⁻² fo ancymidol and daminozide, 200 ml·m⁻² for uniconazole, and 200-300 ml·m⁻² for chlormequat and paclobutracol.

elongation of *Lilium longiflorum* Thunb. 'Nellie White' depended on the light intensity (Wilkins et al., 1986). Thus, applications must be adjusted based on prevailing environmental conditions and the level of desired control. In general, higher rates of growth retardants are used when cultural conditions favor rapid growth or when greater, more long-lasting control is desired. Higher rates are recommended for more tropical climates, and lower rates for more temperate climates. The A-Rest (ancymidol) label recommends increasing the standard rate by 50 percent for use on finished bedding plants that a grower wishes to hold for several weeks. Conversely, low rates are used when weather conditions do not favor rapid growth or when only a short-term effect is desired.

Species Sensitivity to Growth Retardants

The recommended application rates vary among species. A particular concentration of chemical may result in little or no reduction in stem elongation, or extreme and prolonged stunting of plant growth, depending on plant species. For example, daminozide was effective as a growth retardant on a number of crops, including *Calendula officinalis* L. 'Mandarin' (Armitage, Bergmann, and Bell, 1987), various *Hydrangea macrophylla* Ser. cultivars (Bailey and Weiler, 1984), *Salpiglossis sinuata* R. et P. (Needham and Hammer, 1990), *Dendranthema* 'Always Pink' (Holcomb, Ream, and Reid, 1983), *Euphorbia* 'Annette Hegg Hot Pink' (McAvoy, 1990), and various *Eustoma grandiflorum* Shinn (Whipker, Eddy, and Hammer, 1994a) and *Aster novi-belgii* L. cultivars (Whipker et al., 1995). However, daminozide was relatively ineffective for controlling stem elongation in *Hypoestes phyllostachya* Bak. 'Pink Splash' (Foley and Keever, 1992; Starman and Gibson, 1992) or *Bouvardia humboldtii* Hort. (Wilkinson and Richards, 1987). Paclobutrazol at 0.15 mg or uniconazole at 0.025 mg controlled stem height of container-grown bedding plants as a spray or drench, except for *Zinnia elegans* Jacq. 'Yellow Marvel' (Banco and Stefani, 1988). The concentration of paclobutrazol required to control growth of *Zinnia* was eightfold greater than for *Pelargonium* (Cox and Keever, 1988). On woody shrubs, a 500 ppm spray of flurprimidol reduced shoot growth of many species, except *Euonymus kiautschovica* Loes., which required

1,200 ppm (Banco and Stefani, 1995). *Ligustrum* ×*ibolium* E.F. Coe and *Photinia* ×*fraseri* Dress were more sensitive to uniconazole than *Pyracantha kodzumii* Rehd. (Norcini and Knox, 1989b). A 5 mg dose of paclobutrazol was excessive for *Pyracantha* 'Mohave' or *Juniperus chinensis* L. 'San Jose' (Ruter, 1994), but a drench of 10 mg/plant was suitable for *Buddleja davidii* Franch. 'Dubonnet' (Ruter, 1992). The relative efficacy among the growth-retardant chemicals also varies among species. Uniconazole is generally more effective than paclobutrazol in inhibiting stem growth. However, the response of *Kalmia latifolia* L. 'Carousel' and 'Yankee Doodle' to paclobutrazol was greater than that of *Rhododendron catawbiense* Michx. 'Boursault' and 'Roseum Elegans', and, consequently, these two chemicals had similar effects on *Kalmia* (Gent, 1995). The dose of chlormequat required to control plant height was 30 mg for *Pyracantha coccinea* (Henderson and Nichols, 1991) and 550 mg for *Tibouchina urvilleana* DC. (Roberts, Eaton, and Seywerd, 1990), while uniconazole was effective at a dose of 0.25 mg/plant for both species.

Control of stem elongation is not the only factor governing the commercial utility of a growth retardant. For example, paclobutrazol, ancymidol, daminozide, and chlormequat all control shoot elongation to some extent in *Rhododendron obtusum* Planch 'Gloria', but paclobutrazol resulted in plants with the highest quality (Whealy, Nell, and Barrett, 1988). Both chlormequat and ancymidol drenches reduce stem elongation of *Euphorbia* 'Annette Hegg Diva' (Wilfret, Harbaugh, and Nell, 1978). Ancymidol provided a more effective, longer-lasting control but also reduced bract size, while chlormequat did not have this adverse effect. Ancymidol provided more long-lasting control than daminozide and chlormequat on *Gerbera jamesonii* Hook. 'Mardi Gras' and 'Jongenelen' (Armitage, Hamilton, and Cosgrove, 1984; Armitage, Tu, and Vines, 1984).

Cultivar Response

The response to a specific growth retardant varies not only among species but also among cultivars within a species. For example, a paclobutrazol drench of 10 to 40 ppm reduced the height of *Matthiola incana* L.R. Br. 'Lavender' by 60 to 80 percent and delayed flowering as much as three weeks, but the same treatments had no effect on either height or flowering of the naturally compact cultivar Midget Red (Ecker et al., 1992). Similarly, a 15 ppm spray of paclobutrazol reduced stem elongation of the tall ornamental *Brassica oleracea* L. 'Golden Emblem', but not that of the short cultivar Red Pygmy (Whipker, Eddy, and Hammer, 1994b). Among *Lilium longiflorum* cultivars, a paclobutrazol drench of 4 to 20 ppm controlled stem elongation of the cultivar 'Ace', but not that of 'Nellie White' (Gianfagna and Wulster, 1986). In contrast, an ancymidol drench at 4 to 20 ppm was effective for both Ace and 'Nellie White'. The effect of uniconazole on flowering was dependent on cultivars among *Lilium* hybrids (Jiao, Wang, and Tsujita, 1990). Cultivars of large-leaf *Rhododendron* and *Kalmia* varied in their response to sprays of 4 to 30 ppm paclobutrazol or 1 to 12 ppm uniconazole. Stem elongation of *Rhododendron* 'Nova Zembla' was more responsive than 'Roseum Elegans', which was more responsive than 'Boursault', and *Kalmia* 'Yankee Doodle' was more sensitive than 'Carousel', but the former was less likely to flower in response to treatment (Gent, 1995).

EFFICACY AND METHOD OF APPLICATION

Growth-retardant chemicals are usually applied as a whole-plant spray or as a root-medium drench. Both of these methods have benefits and drawbacks. Sprays are more likely to be phytotoxic. When grown in large multiplant containers, uniformity, both in coverage and ensuing plant response, is difficult to achieve with growth-retardant sprays. Sprays are attractive because they can be applied relatively quickly. Drench applications are more labor intensive but produce a greater, longer-lasting, and more uniform plant response. However, the effect of a drench application on young seedlings, such as "plug" transplants of annual species, may be too dramatic and too long-lasting. Incorporating growth retardant into the potting medium in an encapsulated form before or after planting can also provide an enduring and desirable level of control over plant growth.

Drench

For most growth-retardant chemicals, a root-medium drench is more effective than a whole-plant spray. A root drench of ancymidol was generally more effective than a spray for controlling plant height of various bedding plant species (Murray, Sanderson, and Williams, 1986). Uniconazole at 0.25 mg/plant as a drench, but not as a spray, controlled height in seed-propagated *Pelargonium* 'Multibloom Scarlet' and 'Red Elite' (Starman, Cerny, and Grindstaff, 1994). However, a uniconazole spray was more effective than a drench when applied to *Lilium* hybrids when plant height was 10 centimeters (cm) (Jiao, Wang, and Tsujita, 1990). An ancymidol drench inhibited plant height of *Eustoma* more than desired, but a more concentrated solution was suitable when applied as a spray (Tija and Sheehan, 1986). For various woody landscape species, a drench of uniconazole reduced stem weight more than a spray (Warren, Blazich, and Thetford, 1991). For a variety of ornamental species, a drench application of ancymidol was most effective when roots were actively growing, and less effective when applied to plants with well-established roots (Cathey, 1975).

Spray Adjuvants

All commercial growth retardants contain spray adjuvants, and the use of additional spreader, sticker, or uptake-enhancing adjuvants on ornamental plants is not recommended. However, surfactants may improve efficacy of spray application. The effect of ancymidol on *Hydrangea* 'Merveille' and 'Rose Supreme', and *Lilium longiflorum* 'Georgia' was greater when it was sprayed along with a surfactant (Tija et al., 1976). The efficacy of paclobutrazol to control height of *Dendranthema* cultivars was increased when it was sprayed in a solution of 15 percent ethanol (McDaniel, 1983). Paclobutrazol more effectively reduced growth and leaf area of *Fragaria* ×*Ananassa* Duch. 'Belrubi' when applied with surfactant (Ramina and Tonutti, 1985). When paclobutrazol was applied as a spray to *Malus,* it inhibited shoot growth of 'Delicious' at tight cluster stage only when combined with the surfactants Tween or Triton (El-Khoreiby, Lehman, and Unrath, 1989), and increasing the concentration applied to 'Cox' or 'Suntan' had greater effect with surfactant than without (Richardson, Webster, and Quinlan, 1986).

Bulb Dip or Soak

Soaking bulbs before planting is an alternative to the use of sprays and drenches applied during shoot elongation. Growth of *Lilium* hybrid 'Enchantment' was retarded more by uniconazole if bulbs were soaked for 30 minutes (min) in a 0.25 to 2.0 ppm solution than if the root medium was drenched after shoot emergence (Bearce and Singha, 1990). A bulb soak in 10 ppm ancymidol had similar effects on other *Lilium* hybrids (Parivar, Preece, and Coorts, 1985). Stem height of *Lilium longiflorum* 'Ace' and 'Nellie White' was controlled effectively when bulbs were soaked in a 5 to 33 ppm solution of ancymidol for 30 to 60 min (Larson et al., 1987; Wulster, Gianfagna, and Clarke, 1987). Although similar results were reported when *Tulipa gesneriana* L. 'Paul Richter' bulbs were soaked in paclo-butrazol, a bulb soak with ancymidol was not as effective as a drench (McDaniel, 1990). A dip or soak can also be applied to the cuttings of vegetatively propagated plants before rooting. There was a linear effect of concentration on stem elongation eight weeks later when *Dendranthema* 'Dalvina' cuttings were dipped in 1.3 to 10 ppm uniconazole, while a single spray of up to 10 ppm had no effect (Schuch, 1994). If unrooted cuttings of *Hibiscus* 'Seminole Pink' were soaked in 25 ppm of uniconazole or paclobutrazol, stem growth was inhibited by 75 percent and 25 percent, respectively (Wang and Gregg, 1991). Vacuum infusion is more effective than a simple dip. Vacuum infusion with daminozide at 250 ppm retarded height growth of *Dendranthema* 'Engarde' as much as a dip in 1,000 ppm (Sanderson, Smith, and McGuire, 1994). When root cubes of *Euphorbia* 'Annette Hegg Brilliant Diamond' were soaked in uniconazole for 15 min prior to planting, rates as low as 0.1 ppm were adequate for height control, provided that the root cubes were initially dry (Bearce and Singha, 1992). Well-watered cubes needed 1 ppm to produce the same results, since these cubes had a lower capacity to absorb the uniconazole solution.

Drench Applications to Bark-Based Media

Drench treatments of ancymidol and paclobutrazol are less effective when applied to pine bark-based media than to media without pine bark. Stem elongation of *Dendranthema* 'Bright Golden Anne' grown in a pine-bark medium was not affected by a drench of 0.25 mg/plant ancymidol, flurprimidol, or paclobutrazol, but it was greatly reduced in media without pine bark (Barrett, 1982). A bulb dip in 16 to 33 ppm ancymidol controlled stem elongation of *Lilium longiflorum* 'Ace' and 'Nellie White' grown in a pine-bark medium, whereas a drench did not (Larson et al., 1987). 'Nellie White' failed to respond to ancymidol drench in pine-bark medium, but it did respond when ancymidol-impregnated polyacrilamide gel was incorporated into the medium (McAvoy and Kishbaugh-Schmidt, 1992). A lettuce-hypocotyl bioassay revealed 70 percent of the ancymidol applied as a drench was in the upper third of the medium, while ancymidol delivered in the polyacrilamide gel carrier was distributed uniformly throughout the rooting medium.

Preplant Incorporation of Retardant into the Rooting Medium

Incorporating growth retardant into the potting medium in an encapsulated form before planting can provide an enduring and desirable level of control over plant growth. Microencapsulated chlormequat incorporated into the root medium prior to planting effectively controlled growth of a number of bedding plant species (Read, Herman, and Heng, 1974). Other preplant application methods include impregnating clay pots with 20 to 40 ppm ancymidol (Einert, 1976), or a rock wool-medium amendment with uniconazole (McAvoy, 1990). Acrylamide gel impregnated with 0.018 mg uniconazole and incorporated into potting medium prior to planting controlled the height of *Lilium* 'Ace' as well as an ancymidol drench at 0.5 mg per pot applied after shoot emergence (McAvoy, 1991). Chlormequat applied at a constant low rate of 55 to 100 ppm for 15 to 30 days was as effective as a single drench of 3,000 ppm to control height of *Euphorbia* 'Paul Mikkelsen' (Holcomb and White, 1970). Preplant treatments of daminozide and uniconazole were more effective than postplant treatments to control height of *Dendranthema* 'Bright Golden Anne' (Hicklenton, 1990).

Postplant Incorporation of Retardant into the Rooting Medium

Tablets, capsules, gels, and spikes produce mixed results when they are used to apply growth retardant during plant growth. Height of *Clerodendron thomsoniae* Balf. was controlled by gels soaked with 360 mg chlormequat, or 0.5 mg of either ancymidol or paclobutrazol (Sanderson, 1990; Sanderson, Martin, and McGuire, 1990).

The effect on *Alcea rosea* L. 'Powderpuff Mix' of paclobutrazol at 0.13 mg/plant was similar when applied as impregnated spikes, a spray, or a drench (Latimer, Oetting, and Thomas, 1995). An ancymidol drench at 0.25 mg/plant had the same effect as paclobutrazol-impregnated spikes on the height of *Tulipa* 'Kees Nelis' (Deneke and Keever, 1992). Paclobutrazol-impregnated spikes were also as effective as a drench on *Codiaeum variegatum* L., *Brassaia actinophylla* Endl., *Euphorbia* 'Annette Hegg Dark Red', and *Impatiens wallerana* L. 'Super Elfin Red' and 'Show Stopper', but not on *Caladium* ×*hortulanum* Birdsey 'White Christmas' and 'Carolyn Wharton' (Barrett, Bartuska, and Nell, 1994b). A 0.5 mg ancymidol drench provided better height control of various *Euphorbia* cultivars than an ancymidol-impregnated tablet, and the tablet inhibited bract size more than the drench (Sanderson, Martin, and Reed, 1989). The tablet acted similar to a late-season drench, indicating the distribution and availability in the root zone were affected by the application method. In comparing the effects of 0.5 mg/plant of paclobutrazol applied as tablets, drenches, gels, capsules, and sprays to *Dendranthema* 'Charm', a drench was more effective than other application methods (Sanderson, Martin, and McGuire, 1988), perhaps because the chemical leached slowly from the impregnated carriers. If *Euphorbia* 'Eckespoint Celebrate 2' was treated with paclobutrazol-impregnated spikes or a drench at 1 mg/plant, the number of spikes was not critical, but they had to be applied earlier in the growth cycle than a drench to achieve a similar level of control (Newman and Tant, 1995). However, in bark-based media, spikes impregnated with 0.5 mg/plant paclobutrazol retarded growth in *Lantana camara* L. 'New Gold' more effectively than a drench (Ruter, 1996).

UPTAKE, MOVEMENT, AND PERSISTENCE IN PLANTS

The consensus is that growth retardants, except daminozide, are taken up by roots, or through the bark, and transported in the xylem only toward the leaves and stem apex. This has been shown using ^{14}C-tracers of triazoles applied to *Malus*. When 'York Imperial' seedlings were grown in nutrient solution with paclobutrazol, they took it up through roots and transported it primarily in xylem. After forty-two days of uptake, the distribution of ^{14}C-labeled compound was 53 percent in leaves, 26 percent in stem, and 21 percent in roots (Wang, Sun, and Faust, 1986). One month after ^{14}C-paclobutrazol was injected in the stem, 23 percent of the chemical was translocated to apical shoots, 90 percent in its original form (Sterrett, 1985). One month after ^{14}C-uniconazole was injected into one-year-old trees, more than 96 percent of the label was in xylem and phloem tissue, and 92 percent was still in the form of uniconazole (Sterrett, 1988). Two months after injecting flurprimidol into *Liriodendron tulipifera* L. or *Platanus occidentalis* L. stems, 10 percent of the chemical moved to new shoots, 80 percent was at site of injection, and 90 percent was in the original chemical form (Sterrett and Tworkoski, 1987). When paclobutrazol was applied to individual stems of *Swainsona formosa*, growth was inhibited only on the treated shoot, whereas gibberellin applied the same way enhanced apical dominance throughout the plant (Hamid and Williams, 1997). However, paclobutrazol appeared to be phloem mobile in *Ricinus communis* L., which would imply translocation throughout the plant (Wichard, 1997).

Uptake Rates

The rate of uptake of growth retardant into plants varies among the chemicals. Ancymidol, paclobutrazol, and uniconazole appear to rapidly diffuse into the plant, typically within a few minutes after spray application. Whereas ancymidol activity is maximal within 5 min of a spray application, daminozide activity is adversely affected by overhead irrigation applied hours after treatment (Cathey, 1975). Overhead irrigation applied 30 min after *Dendranthema* 'Nob Hill' was sprayed with ancymidol, paclobutrazol, or uniconazole had no effect on growth-retardant activity, but irrigation prior to 4 h diminished the effectiveness of daminozide (Barrett, Bartuska, and Nell, 1987). Drench applications of ancymidol, paclobutrazol, and uniconazole were not affected by irrigation as early as 1 h after treatment.

Uptake by Stems

In part, sprays are less effective than drenches because of resistance to transport across the cambium layer of the stem. In established trees, paclobutrazol, uniconazole, and flurprimidol were more effective when applied by trunk injection than soil drench (Kimball,

1990). Bark banding was only effective on young trees with thin bark. The variation among trees in resistance to transport across the suberized layer of bark affected the efficacy of growth retardant after trunk painting (Sachs et al., 1986). Uniconazole retarded axillary shoot elongation of *Hibiscus* 'Jane Cowl' when applied directly to the axillary bud or green bark immediately below the bud (Wang, 1991), but it had a limited effect when applied to suberized bark or to leaf blades. In part, sprays are less effective than drenches because any chemical that falls on the leaves is not active as a growth retardant. Uniconazole was an effective growth retardant when applied to *Dendranthema* 'Nob Hill' and 'Tara', as a drench or a whole-plant or stem spray, but it was not effective when applied to leaves (Barrett, Bartuska, and Nell, 1994a). Paclobutrazol most effectively controlled height of *Alcea* 'Powderpuff Mix' when applied in a spike or as a drench, while a fog or electrostatic spray was least effective (Latimer, Oetting, and Thomas, 1995). Paclobutrazol had little effect on stem elongation when painted on fully mature leaves of *Phaseolus vulgaris* L. and *Dendranthema* 'Bright Golden Anne', but was very effective when painted on the stem, applied as a drench to the root, or applied as a spray to the whole shoot (Barrett and Bartuska, 1982).

Metabolism

All growth retardants seem to be metabolized relatively slowly in plants. An early study of metabolism of daminozide implied the decrease in concentration of the chemical in the plant was due to dilution by new biomass rather than catabolism of the chemical (Dicks and Charles-Edwards, 1973). This led to a quantitative description of inhibition by daminozide of stem growth in vegetative lateral shoots of *Dendranthema* 'Bright Golden Anne' (Dicks 1972). Based on weekly measurements of internode length, growth was retarded for twenty-one to twenty-five days with a 0.6 mg uniconazole drench, and for fourteen to seventeen days with a 500 ppm daminozide spray (Tayama and Carver, 1992a, b). Concentrations of 500 and 4,000 ppm of a single foliar application of chlormequat to *Euphorbia* 'Annette Hegg Dark Red' affected the duration of growth-retarding activity as well as the degree of inhibition immediately after application (Fisher, Hiens, and Lieth, 1996). The effectiveness of the various growth-retardant chemicals may be directly related to their persistence in plants. After a soil drench or hydroponic application of the triazoles, paclobutrazol, triapenthanol, or BAS111 to *Malus*, paclo-butrazol persisted the longest in leaves or roots, and it had the greatest effect on stem elongation (Curry and Reed, 1989; Reed, Curry, and Williams, 1989). The triazoles are also broken down relatively slowly in potting media due to microbial degradation (Jackson, Line, and Hasan, 1996). Chlormequat is converted to choline in shoots of *Dendranthema* and *Hordeum vulgare* L. (Schneider, 1967). This may be why chlormequat is less effective at a given dose than the chemicals that are not metabolized.

Persistence

The persistence of their action can be a problem with some of the more effective growth-retardant chemicals. As an extreme example, ten years after application of about 1 gram (g) paclobutrazol per cm diameter at breast height, the biomass trimmed from *Acer rubrum* and *A. saccharinum* trees was still less than that of untreated trees (Burch, Wells, and Kline, 1996). This long-term persistence is also seen in woody ornamental plants. *Ilex crenata* 'Compacta' or *Ilex ×mesereae* S.Y.Hu 'China Girl' outgrew the effects of 500 ppm flurprimidol spray in the year after application, but a 1,500-1,800 ppm spray inhibited growth for at least two years (Laiche, 1988; Keever, Gilliam, and Eakes, 1994). The growth of *Photinia* drenched with up to 250 mg flurprimidol in July and planted in soil in December was reduced in the following year (Laiche, 1988). Plant size and lateral branching of *Photinia* was reduced for three to six months by a spray of paclobutrazol, and for up to twelve months with uniconazole at 90 to 140 ppm (Owings and Newman, 1993). Spray applications of 10 to 30 ppm paclobutrazol or 4 to 12 ppm uniconazole to potted *Rhodo-dendron* or *Kalmia* nearly always inhibited stem elongation in the following year, when they were transplanted into the field (Gent, 1997); and growth of *Kalmia* 'Yankee Doodle' was inhibited for three years after application of 100 ppm paclobutrazol. The effect of growth retardant also persists in annuals and herbaceous perennials, but for a shorter period than in woody ornamentals. Stem elongation of several species of bedding plants was inhibited for four to eight weeks by a 10 ppm spray of uniconazole; *Impatiens* 'Zenith' and *Pelargonium* 'Ringo Deep Scarlet' were inhibited a further five to seven weeks after transplant when sprayed with concentrations greater than 10 ppm (Keever and Foster, 1991a). Whereas 7 ppm paclo-butrazol, 1,500 ppm chlormequat, or 200 ppm ancymidol reduced stem length and shoot weight and time to flower of seed *Pelargonium* 'Ringo White' and 'Ringo Rose' up to four weeks after planting, paclobutrazol concentrations greater than 16 ppm inhibited stem elongation for twelve weeks (Latimer and Baden, 1994). Two sprays of chlormequat at 1,500 ppm reduced plant height of cutting-propagated *Pelargonium* 'Sooner Red' and 'Yours Truly' throughout the season (Schwartz, Payne, and Sites, 1985).

Stage of Plant Development

Plant response is affected by the stage of plant development when growth retardant is applied. When a 5 to 15 ppm spray of uniconazole was applied to *Capsicum annuum* L. 'Holiday Cheer' eight weeks after sowing, stem elongation was inhibited more than by a spray at ten weeks, but treatment at ten weeks increased the percentage of red fruit more than at eight weeks (Starman, 1993). The effect of paclobutrazol on inflorescence length of a *Phalaenopsis amablis* L. hybrid depended on time of application (Wang and Hsu, 1994). It was equally effective at 125 to 500 ppm if applied prior to inflores-cence emergence, but the effectiveness decreased when applied after inflorescence emergence. Growth retardants can change the ratio of peduncle to stem length. If a GA_3 spray was applied three weeks after a uniconazole drench of *Dendranthema* 'Echo', it resulted in the same plant height as untreated plants, but the peduncles were 33 per-cent of total height as compared to 18 percent in untreated controls (Holcomb, Tukey, and Rose, 1991). In general, growth retardants applied within a few days after a pinch are more effective than those applied later. Uniconazole at 50 ppm applied with a brush to the bark of *Hibiscus* 'Jane Cowl' immediately following a pinch controlled elongation of the first three internodes on axillary shoots, but this treatment had little effect when applied twenty-four days after pinch-ing (Wang, 1991). When 60 ppm paclobutrazol or 30 ppm uni-conazole was sprayed on *Dendranthema* 'Bright Golden Anne' at zero, two, or four weeks following a pinch, a greater suppression of stem elongation and a longer time to flowering was correlated with an earlier time of application (Gilbertz, 1992). A spray or drench of 2.5 to 10 mg/plant uniconazole more effectively inhibited growth of the woody shrubs *Pyracantha* 'Wonderberry', *Ligustrum lucidum*

Ait., or *Ligustrum sinense* Lour. 'Varigatum' if they were pruned one day before spray application (Norcini and Knox, 1989a).

PHYTOTOXICITY

A phytotoxic response can be a drawback to application of growth retardant to ornamental plants. Sometimes a repeated application at lower concentrations results in height control, when a single higher dose produced a phytotoxic response. At the concentrations required for adequate height control, 6,000 ppm chlormequat and 170 ppm ancymidol, leaves of *Dianthus* 'Persian Carpet' and 'Indian Carpet' were harmed, but leaves of 'Snowfire' recovered from a lower effective concentration (Messinger and Holcomb, 1986). Chlormequat caused pale leaves and marginal chlorosis in a number of species, unless sprayed at less than 1,500 ppm (Cathey, 1975). Leaf necrosis was seen on *Pilea cadierei* Gag. et Guil. sprayed with more than 500 ppm paclobutrazol—the concentration needed to control stem elongation (Cox and Wittington, 1988)—and the severity of damage increased with concentration. A drench of as little as 0.2 mg paclobutrazol or 0.025 mg uniconazole per plant resulted in extreme and prolonged stunting of *Plectranthus australis* R.Br. (Wang and Blessington, 1990). The labels for Bonzi (paclobutrazol) and Sumagic (uniconazole) warn that *Begonia* ×*semperflorens-cultorum* is extremely sensitive and can be easily stunted by an inadvertent drift of spray from these products. *Pelargonium* ×*hortorum* is also extremely sensitive to drench applications of paclobutrazol. Because the efficacy of these chemicals has not been fully appreciated, some reports of phytotoxicity due to uniconazole and paclobutrazol were due to applications at doses that far exceed those required to partially inhibit stem elongation. At such concentrations, ancymidol, paclobutrazol, and uniconazole generally have fewer phytotoxic side effects than chlormequat and daminozide.

EFFECT ON FLOWERING

A major concern in production of ornamental plants is a delay in flowering. Gibberellin biosynthesis inhibitors affect the rate of flower development in many species. This may be a direct consequence of changes in concentration of gibberellins, or it may be a secondary effect of reducing the competition between vegetative and reproductive parts for sugars and mineral nutrients. The relationship between growth retardants and flowering appears to be complex. Where it has been analyzed in more detail, gibberellin biosynthesis inhibitors have effects on particular transitions during flower development (Pharis and King, 1985). Paclobutrazol applied before flowering at 0.1 mg/plant to *Zinnia* 'Red Sun' delayed the transition to reproductive stage by four days and produced single flowers with ray petals converted to tubular petals (Kim et al., 1989).

Fewer or Delayed Flowers

Growth retardants prevent flowering, reduce the number of flowers, or retard flower development in many ornamental species. Bulb species require gibberellins for flower development. An isolated floral stalk of *Tulipa* 'Apeldoorn' treated with paclobutrazol produced a fully developed flower only with addition of gibberellic acid (Rebers, Romeijn, and Knegt, 1994). Thus, gibberellin biosynthesis inhibitors often delay flowering of bulbs. A paclobutrazol or ancymidol drench at 0.25 to 0.75 mg/plant delayed flowering in *Tulipa* 'Paul Richter' (McDaniel, 1990). Flowering of *Lilium* 'Nellie White' was delayed up to three days by a 50 ppm uniconazole spray (Bailey and Miller, 1989a, b), although 10 ppm was sufficient to control height. The height of *Lilium* hybrid 'Enchantment' was controlled by 0.1 mg uniconazole or 0.2 mg ancymidol as a drench, but these treatments also delayed flowering (Bearce and Singha, 1990). Ancymidol delayed flowering in *Freesia hybrida* Bailey cultivars grown from unvernalized corms, but not when corms were vernalized prior to forcing (Wulster and Gianfagna, 1991). *Zantedeschia rehmannii* Engl. rhizomes soaked to take up 2 to 4 mg paclobutrazol produced shorter plants with fewer flowers than did untreated rhizomes (Corr and Widmer, 1991). Sprays of 0.5 mg ancymidol or 5,000 ppm daminozide when flower buds were at canopy height reduced peduncle length of potted *Ranunculus asiaticus* L. 'Tecolote Giant White' without delaying anthesis, but preplant tuber dips in 10 to 25 ppm ancymidol or 3 to 10 ppm flurprimidol caused excessive stunting and delayed flowering (Albrecht, 1987). A daminozide dip delayed flowering of *Phalaenopsis* by five to thirteen days, but inflorescence height was not affected (Wang and Hsu, 1994), whereas a dip in

400 ppm paclobutrazol or 200 ppm uniconazole reduced plant height without delaying flowering. Thus, the stage of development when growth retardants are applied, as well as the chemical used, affects the delay in flowering.

Growth retardant can also delay flowering of nonbulb annual and perennial species. Flowering in *Rudbeckia hirta* 'Marmalade' was delayed by a 50 ppm ancymidol drench (Orvos and Lyons, 1989), and height was more effectively controlled by limiting the number of long days used to induce flowering than by ancymidol. Neither dikegulac nor chlormequat affected height or quality of *Begonia* ×*hiemalis* Fotsch. 'Northern Sunset', but dikegulac plus hand pinch reduced flower count (Agnew and Campbell, 1983). In the range from 0.25 to 2 mg/plant, a drench of paclobutrazol delayed the days to flower of the tall *Matthiola* 'Lavender', but not the shorter 'Midget Red' (Ecker et al., 1992). A 1.2 to 1.8 mg/plant drench of paclobutrazol reduced plant height, increased the number of flower spikes, and delayed flowering of *Antirrhinum majus* L. 'Coronette' (Wainwright and Irwin, 1987). Some woody species are also adversely affected. A uniconazole drench of up to 0.4 mg/plant increased time to flower in *Hibiscus* 'Jane Cowl' (Wang and Gregg, 1989). Flowering of *Hydrangea* 'Rose Supreme' was delayed twelve days or more by a 2 × 10 ppm spray of uniconazole (Bailey, 1989a). A similar effect was observed when daminozide was applied to *Hydrangea* forty-two days before flower primordia become visible (Bailey and Weiler, 1984).

The delay in flowering can be due to applying an excessive dose of growth retardant. Uniconazole decreased the flowering of *Pharbitis nil*, in a manner related to the decrease in endogenous gibberellins (Wijayanti et al., 1996). Flower development of *Rhododendron obtusum* 'Gloria' was delayed by sprays of daminozide and chlormequat, but not by ancymidol or paclobutrazol (Whealy, Nell, and Barrett, 1988). However, higher concentrations of paclobutrazol, 50 to 150 ppm, delayed flowering three to eight days (Keever, 1990), and 200 ppm uniconazole delayed flowering up to eighteen days (Keever and Foster, 1991b). A chlormequat drench applied to *Pelargonium* 'Red Elite' at the sixth-leaf stage controlled height but did not affect flowering of plants grown in a controlled environment (White and Warrington, 1984a), although the same treatment delayed flowering in a greenhouse (White and Warrington, 1984b). Flowering in *Dendranthema* 'Neptune' was progressively delayed with increasing concentrations of ancymidol drench (Johnson, 1974). Thus, concentrations of growth retardant greater than that required to partly inhibit stem elongation are more likely to affect flower development. Because of the efficacy of the triazole chemicals, the rates used in some studies may have been greater than required for moderate control of stem elongation.

No Effect on Flowering

In many cases, growth retardants control height with a minimal effect on flowering. A daminozide spray at 3,500 ppm applied three weeks after sowing had no effect on time to flower of *Calendula* 'Mandarin' (Armitage, Bergmann, and Bell, 1987). The time from the beginning of short days to flower was not affected by a 2 mg/plant uniconazole spray on various *Euphorbia* cultivars, nor by a 4 mg/plant uniconazole drench on *Dendranthema* 'Ovaro' (Bailey, 1989b). A 10 ppm uniconazole spray had no effect on time of flowering in *Hibiscus* 'Brilliant' (Newman, Tenney, and Follett, 1989). In *Eustoma* 'Yodel Blue', flowering was delayed by a uniconazole drench at 1.6 mg/plant but not by a 10 ppm spray (Starman, 1991). A paclobutrazol drench of 0.2 mg/plant had no effect on time of flowering in *Episcia cupreata* Hanst. 'Pink Panther' (Stamps and Henny, 1986), but it increased flower longevity. The time to flower in vegetatively propagated *Pelargonium* 'Yours Truly' was not affected by 15 ppm paclobutrazol, 10 ppm uniconazole, or 1,500 ppm chlormequat (Tayama and Carver, 1990). A 100 to 200 ppm paclobutrazol spray suppressed shoot elongation of *Dianthus* up to fourteen weeks, without delaying time to flower (Foley and Keever, 1991).

Earlier or More Flowers

Growth retardants advance flowering and increase the number of flowers in several woody species. Stuart (1961) showed that 200 to 400 mg/plant chlormequat, in combination with a day length and cooling treatment, promoted flowering of *Rhododendron obtusum* 'Coral Bells' within a year of propagation. When grown under field conditions, 10 to 30 ppm paclobutrazol or 4 to 10 ppm uniconazole applied in spring to *Rhododendron* 'Boursault' and 'Roseum Elegans' and *Kalmia* 'Carousel' and 'Yankee Doodle' promoted flowering in the following year (Gent, 1995). A 1 to 7 mg drench of paclobutrazol applied after the first flush of *Rhododendron catawbiense* 'Roseum Elegans' also increased flower number (Ranney et al., 1994). An excessive dose of 50 to 400 mg paclobutrazol as a drench even caused multiple flower buds to form on each stem apex of *Rhododendron* hybrid 'Sir Robert Peel' (Wilkinson and Richards, 1991). A 15 to 30 ppm uniconazole spray during the vegetative stage increased flower number and earliness of *Rhododendron* hybrid 'Prize' more than a spray during flower development (Keever and Olive, 1994). A spray of 150 ppm paclobutrazol increased flower number and reduced time to flower more than 5,000 ppm daminozide (Keever and Foster, 1989). Sprays of 6,000 ppm daminozide and 1,500 ppm chlormequat, but not 130 ppm ancymidol or 200 ppm paclobutrazol, slowed development in *Rhododendron obtusum* 'Gloria' (Whealy, Nell, and Barrett, 1988). Although under long days *Bouvardia humboldtii* grew vegetatively, an ancymidol drench induced plants to initiate flowers at the same time as plants under short days (Cathey, 1975). Flowering was promoted by a 5 ppm uniconazole spray on *Camellia* 'Shishi-Gashira' (Keever and McGuire, 1991). Flower initiation in *Hydrangea* 'Merritt's Supreme' was advanced when paclobutrazol was supplied in hydroponic solution at 0.2 ppm for four weeks before the onset of flower initiation (Wilkinson and Hanger, 1992). A drench of paclobutrazol promoted flowering in *Euonymus japonica* 'Microphylla', *Juniperus compacta* Paul. 'Blue Pacific', and *Rhododendron obtusum* 'Hino Crimson' (Keever, Foster, and Stephenson, 1990). Both 130 ppm ancymidol and 1,000 ppm chlormequat stimulated early flower bud initiation and increased the number of flower buds in *Schlumbergera truncata* Haw. (Ho, Sanderson, and Williams, 1985). In seed-propagated *Pelargonium* 'Sprinter Scarlet', a 1,500 ppm chlormequat spray, or a 0.06 mg/plant paclobutrazol drench, accelerated flowering when applied prior to flower initiation, but not when applied after initiation (Armitage, 1986). A 200 ppm ancymidol spray induced flower initiation one to two weeks earlier than in untreated plants (Miranda and Carlson, 1980). Thus, the flowering response to growth retardants is complicated, depending on species, as well as stage of development and environmental factors.

Root Growth

Growth retardants can affect other aspects of plant anatomy, such as root growth. Root thickening is an early morphological response to growth retardant (Avidan and Erez, 1995). Six *Citrus* cultivars treated with paclobutrazol had shorter thicker roots (Yelenosky, Vu, and Wutscher, 1995). Paclobutrazol increased root tip diameter and radial elongation of the inner layer of cortical cells of *Prunus persica* Batsch. 'Redhaven' (Williamson, Coston, and Grimes, 1986). Paclobutrazol also increased root diameter and number of rows and diameter of cortical cells of *Dendranthema* 'Lillian Hoek', but it decreased root vascular development (Burrows, Boag, and Stewart, 1992). Growth retardant often decreases root growth. However, the root:shoot ratio may increase due to a greater reduction in shoot growth than in root growth, as seen in *Prunus* (Early and Martin, 1988; Rieger and Scalabrelli, 1990; Avidan and Erez, 1995) and *Psuedotsuga taxifolia* Britt. and *Pinus taeda* L. (Wheeler, 1987) treated with paclobutrazol. However, a paclobutrazol treatment that reduced height of *Bouvardia* by 30 percent also decreased the root:shoot ratio (Wilkinson and Richards, 1987). The various growth-retardant chemicals may have different effects on root and shoot growth, depending on their effect on gibberellin or sterol metabolism. In *Prunus* 'Rutger's Red Leaf', GA_3 inhibited root growth, so gibberellin biosynthesis inhibitors may also enhance root growth by decreasing gibberellins (Tagliavini and Looney, 1991). In *Triticum aestivum* L., the (2S,3S) paclobutrazol enantiomer that specifically reduced GA_1 inhibited shoot growth more than root growth, while the (2R,3R) enantiomer that inhibited sterol synthesis inhibited root growth (Lenton, Appleford, and Templesmith, 1994). The root:shoot ratio may also be altered due to less demand by the shoot for photosynthate. A 500 ppm paclobutrazol spray decreased ^{14}C uptake by 30 to 50 percent in *Malus* seedlings, and a greater proportion of ^{14}C moved to the lower stem and roots (El Hodairi, Canham, and Buckley, 1988). Ancymidol, chlormequat, or paclobutrazol supplied to *Malus* in nutrient solution increased root weight and polyamine content (Wang and Faust, 1986).

Synthetic auxins are commonly used to induce root formation in commercial propagation of many woody and herbaceous perennial species (Hartmann et al., 1997). Growth retardants can also increase adventitious root formation in cuttings. Rooting was dramatically increased when *Phaseolus* and *Plectranthus* cuttings were soaked for 24 to 40 h in 3 to 6 ppm paclobutrazol (Davis et al., 1985). Both uniconazole and paclobutrazol at 1 ppm in solution increased adventitious rooting on juvenile cuttings of *Hedera* if an auxin was also used (Geneve, 1990). The growth retardants had no effect on rooting of mature tissue or when auxin was not used. The increase in adventitious roots formed on *Phaseolus* hypocotyl cuttings treated with paclobutrazol was associated with increased activity of the enzymes catalase, peroxidase, and especially malate dehydrogenase (Upadhyaya, Davis, and Sankhla, 1986). Paclobutrazol altered rhizome bud formation of in vitro cultures of *Lapageria rosea* Ruiz et Pav. 'Nashcourt' with respect to planes of cell division and expansion (McKinless and Alderson, 1991).

Leaf Area and Thickness

Leaf size may be reduced by a high dose of growth retardants, but not necessarily by a low dose. A 10 ppm ancymidol spray was sufficient to reduce stem length of *Lilium* 'Nellie White', but at more than 50 ppm, it reduced leaf area (Bailey and Miller, 1989b). A modified inverse function predicted leaf elongation of *Cyperus rotundus* L. as a function of concentration of paclobutrazol (Kawabata and DeFrank, 1994). A high dose of 2 to 6 mg/plant uniconazole applied as a spray or a drench reduced leaf number, area, and weight of *Vinca major* (Fuller and Zajicek, 1995). The dose of growth retardant required to control stem elongation on woody plants may also reduce leaf area. Leaf expansion of *Prunus* 'Neemaguard' was more sensitive than stem elongation to paclobutrazol in nutrient solution (Early and Martin, 1988). A drench of 0.5 mg/plant or more of paclobutrazol reduced leaf number and leaf size of *Ficus benjamina* L., and this effect persisted for six months (LeCain, Schekel, and Wample, 1986). A uniconazole drench of up to 3 mg/plant decreased height and leaf area of *Pyracantha* 'Lavender', *Photinia*, and *Ilex cornuta* Lidl. 'Burfodii Nana' (Frymire and Cole, 1992). However, a dose of paclobutrazol that decreased leaf area of *Ligustrum japonica* Thunb. by 50 percent had little effect on leaf area of *Feijoa sellowiana* Berg. (Martin et al., 1994). A drench of 15 mg flurprimidol or 2.5 mg uniconazole decreased plant weight but did not affect leaf area of *Forsythia intermedia* Spaeth 'Spectabilis' (Viagro-Wolff and Warmund, 1987), and leaf area of *Rhododendron* was not affected by low doses of growth retardants. When applied to *Epipremnum aureum* Bunt., a 5 mg paclobutrazol drench optimized leaf size (Conover and Satterthwaite, 1996), and 0.4 mg uniconazole drench increased leaf size and the number of basal lateral shoots (Wang, Hsiao, and Gregg, 1992), while 0.1 mg paclobutrazol or 0.05 mg uniconazole increased the area of leaves expanded after treatment (Wang and Gregg, 1994). A 250 ppm spray of dikegulac increased shoot number and leaf area of *Nephrolepis exaltata* L., with no effect on leaf length (Carter, Singh, and Whitehead, 1996). When leaf area is reduced, it is often accompanied by an increase in thickness (Tezuka, Takahara, and Yamamoto, 1989; Benton and Cobb, 1995), due to development of another layer of spongy mesophyll cells. Although chlorophyll and carotenoid contents increased in *Triticum* leaves when treated with triazole, this was due to smaller, thicker leaves, and not due to increased synthesis of pigments (Khalil, 1995).

Dimensions of Phloem and Xylem

In general, stem diameter does not increase when stem elongation is inhibited by growth retardants. In part, this is because growth retardants decrease the diameter of xylem and phloem cells. Growth retardants reduced secondary cell wall development in phloem cells of *Euphorbia* 'Annette Hegg Dark Red' (McDaniel, Graham, and Maleug, 1990). Cell wall thickening was completely eliminated by 25 ppm paclobutrazol, reduced by chlormequat and ancymidol, and only limited thickening occurred with 10 ppm uniconazole. A 0.4 mg dose of uniconazole decreased secondary xylem diameter, and number and size of phloem fibers, resulting in a cascading growth habit of *Hibiscus* 'Jane Cowl' (Wang and Gregg, 1989). A dose of 0.5 mg paclobutrazol resulted in narrower stems but increased development of secondary xylem of *Dendranthema* 'Lillian Hoek' (Burrows, Boag, and Stewart, 1992). A 90 to 210 ppm uniconazole spray decreased the cross section of xylem and phloem of *Forsythia* 'Spectabilis' (Thetford et al., 1995). Prohexadione applied to bark of the terminal shoot of *Pinus sylvestris* L. inhibited phloem and xylem production in current- and previous-year wood (Wang et al., 1995). Consequently, weak or "rubbery" stems sometimes result from applying growth retardants.

PLANT PHYSIOLOGICAL RESPONSES

Photosynthesis and Carbohydrate Content

Especially at higher doses, growth retardants can affect physiological processes in the plant. Leaves that develop after growth-retardant application tend to be smaller but thicker, and chlorophyll, density of stomata, and specific photosynthesis are often increased per unit area. Photosynthesis and transpiration rates, and chlorophyll concentration of *Pelargonium* 'Scarlet Sprinter', were increased by a 3,000 ppm spray of chlormequat two to three days after treatment and remained high for four to five days (Armitage, Tu, and Vines, 1984). A 90 to 210 ppm uniconazole spray increased stomata density, specific photosynthesis, and conductance of leaves developing after treatment on *Forsythia* 'Spectabilis' (Thetford et al., 1995). However, an extreme dose of 500 mg of paclobutrazol reduced biomass by 80 percent and decreased leaf photosynthesis, carbohydrate content, and dark respiration by 70 to 80 percent in *Citrus sinensis* Osbeck. (Vu and Yelenosky, 1992). Growth retardants decrease specific photosynthesis (per unit area) of leaves that are fully expanded at time of treatment, but they may increase photosynthesis of leaves that develop after treatment. Application of BAS 111 or triapenthanol reduced specific photosynthesis and leaf conductance of fully expanded leaves of *Brassica napus* L. for two weeks, but leaves expanding after application were not affected (Butler et al., 1989). In leaves of *Vitis* 'Seyval Blanc' formed before treatment, a paclobutrazol drench reduced photosynthesis within five days (Hunter and Proctor, 1994). However, several months after paclobutrazol was applied to established *Prunus* 'Fantasia' trees, it decreased leaf area by 55 percent, but not specific photosynthesis or yield (DeJong and Doyle, 1984). Paclobutrazol decreased leaf area and chlorophyll concentration of *Carya illinoensis* Wangh. but slightly increased photosynthesis per unit area (Wood, 1984). One effect of growth retardant may be to reduce demand for photosynthate, which secondarily limits photosynthesis due to feedback inhibition, a process that has been demonstrated in agronomic crop species. Growth retardant may also change leaf photosynthesis by affecting the membrane composition and function of chloroplasts. Ancymidol applied as a 66 to 132 ppm spray to *Helianthus annuus* 'Mammoth Russian' decreased the concentration per unit area of chlorophyll and three xanthophylls, but not that of carotene (Starman, Kelly, and Pemberton, 1989). Epoxiconazole lowered the concentration of sterols in leaves of *Galium aparine* L. by about 80 percent and reduced photosynthetic oxygen evolution by 37 percent (Benton and Cobb, 1997). Such changes may affect electron transfer or membrane integrity and thus have a direct effect on leaf photosynthesis. Growth retardants tend to decrease biomass accumulation on a whole-plant basis.

Growth retardants can alter the sugar and starch content of plants. When flurprimidol inhibited growth of *Myriophyllum spicatum* L., shoot and root dry weight were unaffected by treatment, but treated plants had up to 68 percent more nonstructural carbohydrate (Nelson, 1996). Paclobutrazol-reduced leaf area and chlorophyll concentration in *Carya* but increased total carbohydrates per unit area (Wood, 1984). Paclobutrazol-treated *Oryza sativa* L. used less carbohydrate for shoot growth and stored more starch in crowns, although the starch-degrading activity was not altered (Yim, Kwon, and Bayer, 1997). The most likely cause of this accumulation of carbohydrate is decreased growth, reducing the demand for the products of photosynthesis. However, leaf carbohydrate concentrations were decreased in *Lilium* 'Nellie White' by uniconazole (Bailey and Miller, 1989b), and in *Pelargonium* 'Red Elite' by chlormequat (White and Warrington, 1984b). The effects of growth retardant on nutrient concentration in leaves show no consistent trend, and these may be secondary effects of carbohydrate accumulation or reduced transpiration. Paclobutrazol in solution reduced leaf nitrogen (N), phosphorus (P), and potassium (K) concentrations in *Prunus* 'Neemaguard' rootstock, while calcium (Ca) and magnesium (Mg) were increased (Rieger and Scalabrelli, 1990). However, six *Citrus* cultivars treated with paclobutrazol soil drench had higher concentrations of N, Ca, boron (B), iron (Fe), and manganese (Mn) in leaves (Yelenosky, Vu, and Wutscher, 1995).

Transpiration and Water Use

Growth retardants alter the sensitivity of plants to a number of environmental stresses, such as drought. They can be used to harden off transplants to increase survival under field conditions. Growth retardants prevent water stress by decreasing whole-plant water use, either due to smaller leaves, smaller leaf area, or fewer or smaller stomata in the leaves. Uniconazole at 3 mg/plant decreased by 33 percent the leaf area, number of stomata, transpiration, and water use of *Hibiscus* 'Ross Estey' (Steinberg, Zajicek, and McFarland, 1991b). In combination with antitranspirant, a 150 mg/plant chlormequat spray at bud set reduced transpiration and leaf conductance of *Hydrangea* 'Improved Merveille' at flowering (McDaniel, 1985). Paclobutrazol at 10 micromolar (μM) concentration in growth medium increased leaf wax and reduced leaf dehydration of *Dendranthema* 'Snowden' and *Beta vulgaris* L. (Ritchie, Short, and Davey, 1991). A

2 to 4 mg/plant dose of uniconazole reduced leaf weight and water use of *Vinca major* by 35 percent, and stomatal conductance was only 20 to 50 percent of control plants (Fuller and Zajicek, 1995). However, a 90 to 210 ppm uniconazole spray decreased stomata length but increased stomata density and conductance per unit leaf area of *Forsythia* 'Spectabilis' (Thetford et al., 1995). At 0.5 mg/plant, uniconazole did not affect leaf conductance, water potential, or transpiration rate of *Pyracantha* (Frymire and Henderson-Cole, 1992). Although 3 mg/plant uniconazole reduced leaf area of *Ligustrum japonicum* 'Texanum' by 63 percent, water use per unit leaf area was similar to untreated plants (Steinberg, Zajicek, and McFarland, 1991a). Some reports also note a difference in plant water potential, which may be independent of the effects on stomata. The xylem pressure of *Forsythia* 'Spectabilis' under drought was increased by 15 mg flurprimidol or 2.5 mg uniconazole, but these treatments did not affect leaf area or transpiration (Viagro-Wolff and Warmund, 1987). Paclobutrazol reduced whole-plant water use, but not diffusive resistance of leaves of *Helianthus*, an effect attributed to the change in plant architecture and self-shading caused by reduced internode elongation (Wample and Culver, 1983).

Stress Tolerance

Growth retardants can ameliorate plant response to environmental stresses of temperature extremes, pollutants, and pesticides. Triadimefon has been called a plant multiprotectant because of its ability to counteract stress as well as suppress fungal attack (Fletcher and Hofstra, 1985). No studies have been done of this combined effect in ornamental plants. Chlormequat affected the severity of bacterial diseases on *Hibiscus* (Chase, Osborne, and Yuen, 1987). The commercial product label for B-Nine (daminozide) lists an application for reducing injury from SO_2 and O_3 on *Petunia hybrida*. Ancymidol reduced injury from SO_2 and O_3 on *Celosia argentea*, *Impatiens wallerana*, *Dendranthema*, *Tagetes* sp., and *Zinnia* (Cathey, 1975). A 50 ppm paclobutrazol spray on *Salvia spendens* F. Sellow four days prior to application of carbamate pesticide reduced the phytotoxicity of the pesticide (Latimer and Oetting, 1994). The paclobutrazol-induced tolerance of *Triticum* 'Frederick' and 'Glenlea' leaves to paraquat may involve an increase in antioxidant enzyme activity (Kraus, McKersie, and Fletcher,

1995). Uniconazole seed treatments induced cadmium tolerance in *Triticum* 'Frederick' and decreased solute leakage due to cadmium (Cd) (Singh, 1993). However, a spray of flurprimidol or mefluidide on *Cirsium arvense* L. enhanced shoot, but not root, injury due to herbicides (Tworkoski and Sterrett, 1992). In some cases, growth retardants alone can reduce insect infestation. Mepiquat reduced silverleaf whitefly infestation on *Gossypium* two weeks after application (Flint et al., 1996). Chlormequat reduced spider mite infestations of *Hibiscus* up to six months after application (Osborne and Chase, 1990). However, paclobutrazol did not affect feeding intensity of a weevil on the weed *Eichhornia crassipes* (Van and Center, 1994).

In agronomy, seeds are imbibed with growth retardant to make the seedlings more tolerant to heat and cold stress. Tolerance to heat and cold stress may be induced by stabilizing membranes and preventing the leakiness that usually follows a short-term temperature stress. An abscisic acid analog in combination with tetcyclacis had additive effects on increased chilling resistance in *Oryza* that involved stomatal closure and membrane stabilization (Flores-Nimedez, Dorffling, and Vergara, 1992). Either paclobutrazol or uniconazole sprayed on *Capsicum* at bloom, four to eight weeks prior to fruit harvest, alleviated chilling injury of the fruit during storage at 2°C (Lurie, Ronen, and Aloni, 1995). *Zea mays* L. 'Pioneer 3902' seeds imbibed in paclobutrazol and ancymidol were protected from heat stress and retained photosynthetic efficiency and membrane integrity (Pinhero and Fletcher, 1994). *Triticum* 'Frederick' seeds imbibed in uniconazole had twice the chlorophyll per unit area and increased protection against drought and high-temperature stress (Fletcher and Hofstra, 1990). Paclobutrazol was more effective than propiconazole and tetraconazole in protecting seedlings of *Triticum* 'Katepwa' from heat and drought stress (Gilley and Fletcher, 1997). Germination of *Glycine max* Merr. seeds in uniconazole prevented high-temperature lipid peroxidation by elevating several antioxidants in roots (Upadhyaya et al., 1990). However, paclobutrazol, which prevented some chilling injury and loss of fresh weight and membrane lipids in *Cucumis sativus* L. 'Victory' due to 5°C temperature, had no effect on fatty acid or lipid composition at 12°C (Whitaker and Wang, 1987).

Nongibberellin Metabolism

The gibberellin biosynthesis inhibitors can affect nongibberellin metabolism in plants. These effects may be due to inhibition of cytochrome oxidase enzymes that are involved in metabolic pathways other than gibberellin metabolism. Triazoles inhibit ethylene production in the step from 1-aminocyclopropane-1-carboxylic acid (ACC) to ethylene. Uniconazole inhibited ethylene production due to heat stress by 32 percent, and due to herbicide, by 48 percent, but it did not inhibit ACC synthesis in *Triticum* or *Glycine* seedlings (Kraus, Murr, and Fletcher, 1991). In *Oryza*, the increase in leaf elongation due to ethylene was blocked with uniconazole (Furukawa et al., 1997). Uniconazole reduced ethylene and ACC in *Phaseolus aureus* L. seedlings (Hofstra, Krieg, and Fletcher, 1989). BAS111 inhibited ethylene and elevated ACC production in *Helianthus* cell suspensions (Grossmann et al., 1993). A triazole growth retardant inhibited growth and ethylene production and ACC oxidase in white leaf-base tissue of *Ananas comosus* Merr. (Min and Bartholemew, 1996). Growth retardants may affect the action of endogenous cytokinins. Paclobutrazol enhanced the effect of 6-benzyladenine on adventitious shoot proliferation in *Spathiphyllum floribundum* Schott. but did not behave like cytokinin alone (Werbouck and Debergh, 1996). Paclobutrazol also affected the metabolism of zeatin derivatives in *Spathiphyllum floribundum* and *Anthurium andreanum* Schott. (Werbouck et al., 1996). Paclobutrazol inhibited abscisic acid biosynthesis in *Cercospora rosicola* (Norman, Bennett, and Poling, 1986). Triazoles can affect other enzyme activities. Epiconazole stimulated the antifungal hydrolases, chitinase, and β-1,3-glucanase in *Triticum* 'Star' (Siefert et al., 1996). Paclobutrazol increased catalase, peroxidase, and especially malate dehydrogenase in *Phaseolus* hypocotyl cuttings (Upadhyaya, Davis, and Sankhla, 1986). The inhibition of translocation of carbon from leaves to other organs in *Zinnia* 'Red Sun' due to uniconazole was correlated with decreased activity of soluble acid invertase (Kim and Suzuki, 1989). Tetcyclacis applied to the herb *Trigonella foenum-graecum* L. increased the concentrations of 14α-methyl sterols, such as cholesterol, from 4 to 38 percent and decreased sapogenin in root tissue (Cerdon et al., 1995). Daminozide increased the root thiocyanate content of *Raphanus sativus* L. 'Burpee White' and *Brassica rapa* L. 'Snow Ball' as seedlings in the greenhouse and as mature plants in the field (Chong, Kanakis, and Bible, 1982). Chlormequat increased artemisinin yield in *Artemesia annua* L. and decreased plant height (Shukla et al., 1992). Ethephon and daminozide influenced essential oil content of *Mentha piperita* L. and *Salvia officinalis* L. (El-Keltawi and Croteau, 1986).

FUTURE TRENDS

A recent trend in applied research is the use of a cocktail of growth retardants to control height and reduce adverse side effects. Two chemicals are used that have different modes of action, or that affect different steps in gibberellin biosynthesis. For instance, the combination of daminozide plus chlormequat has been used on *Euphorbia pulcherrima*. Chlormequat is applied at the highest rate that does not cause injury, and the concentration of daminozide is increased until the desired plant height is achieved. This combination is in experimental use on a variety of herbaceous ornamentals. More recent trials with mixtures of daminozide plus paclobutrazol appear promising for a wide range of ornamental crops.

REFERENCES

Agnew, N.H. and R.W. Campbell (1983). Growth of *Begonia* ×*hiemalis* as influenced by hand-pinching, dikegulac, and chlormequat. *HortScience* 18:201-202.

Albrecht, M.L. (1987). Growth retardant use with potted anemone and ranunculus. *Journal of the American Society for Horticultural Science* 112:82-85.

Armitage, A.M. (1986). Chlormequat-induced early flowering of hybrid geranium: The influence of gibberellic acid. *HortScience* 21:116-118.

Armitage, A.M., B. Bergmann, and E.L. Bell (1987). Effect of daminozide and light intensity on growth and flowering of calendula as a potted plant. *HortScience* 22:611-612.

Armitage, A.M., B.M. Hamilton, and D. Cosgrove (1984). The influence of growth regulators on gerbera daisy. *Journal of the American Society for Horticultural Science* 109:629-632.

Armitage, A.M., Z.P. Tu, and H.M. Vines (1984). The influence of chlormequat and daminozide on net photosynthesis, transpiration, and photorespiration of hybrid geranium. *HortScience* 19:705-707.

Arron, G.P., S. de Becker, H.A. Stubbs, and E.W. Szeto (1997). An evaluation of the efficacy of tree growth regulators paclobutrazol, flurprimidol, dikegulac and uniconazole for utility line clearance. *Journal of Arboriculture* 23:8-15.

Arzee, T.H., H. Langenauer, and J. Gressel (1977). Effects of dikegulac, a new growth regulator, on apical growth and development of three compositae. *Botanical Gazette* 138:18-28.

Avidan, B. and A. Erez (1995). Studies of the response of peach and nectarine plants to gibberellin biosynthesis inhibitors in a hydroponic system. *Plant Growth Regulation* 17:73-80.

Bailey, D.A. (1989a). Uniconazole effects on forcing of florists' hydrangeas. *HortScience* 24:518.

Bailey, D.A. (1989b). Uniconazole efficacy on chrysanthemum and poinsettia is not affected by spray carrier volume. *HortScience* 24:964-966.

Bailey, D.A. and W.B. Miller (1989a). Response of oriental hybrid lilies to ancymidol and uniconazole. *HortScience* 24:519.

Bailey, D.A. and W.B. Miller (1989b). Whole-plant response of Easter lilies to ancymidol and uniconazole. *Journal of the American Society for Horticultural Science* 114:393-396.

Bailey, D.A. and T.C. Weiler (1984). Stimulation of inflorescence expansion in florists hydrangea. *Journal of the American Society for Horticultural Science* 109:792-795.

Banco, T.J. and M.A. Stefani (1988). Growth response of selected container-grown bedding plants to paclobutrazol, uniconazole, and daminozide. *Journal of Environmental Horticulture* 6:124-129.

Banco, T.J. and M.A. Stefani (1995). Cutless and atrimmec for controlling growth of woody landscape plants in containers. *Journal of Environmental Horticulture* 13:22-26.

Banco, T.J. and M.A. Stefani (1996). Growth response of large established shrubs to cutless and atrimmec trim-cut. *Journal of Environmental Horticulture* 14:177-181.

Barrett, J.E. (1982). Chrysanthemum height control by ancymidol, PP333, and EL500 dependent on medium composition. *HortScience* 17:896-897.

Barrett, J.E. and C.A. Bartuska (1982). PP333 effects on stem elongation dependent on site of application. *HortScience* 17:737-738.

Barrett, J.E., C.A. Bartuska, and T.A. Nell (1987). Efficacy of ancymidol, daminozide, flurprimidol, paclobutrazol, and XE-1019 when followed by irrigation. *HortScience* 22:1287-1289.

Barrett, J.E., C.A. Bartuska, and T.A. Nell (1994a). Application techniques alter uniconazole efficacy in chrysanthemums. *HortScience* 29:893-895.

Barrett, J.E., C.A. Bartuska, and T.A. Nell (1994b). Comparison of paclobutrazol drench and spike applications for height control of potted floriculture crops. *HortScience* 29:180-182.

Barrett, J.E. and T.A. Nell (1989). Comparison of paclobutrazol and uniconazole on floriculture crops. *Acta Horticulturae* 251:275-280.

Barrett, J.E. and T.A. Nell (1990). Factors affecting efficacy of paclobutrazol and uniconazole on petunia and chrysanthemum. *Acta Horticulturae* 272:229-234.

Bearce, B.C. and S. Singha (1990). Growth and flowering response of Asiatic hybrid lilies to uniconazole. *HortScience* 25:1307.

Bearce, B.C. and S. Singha (1992). Response of poinsettia to preplant root-zone soaks in uniconazole. *HortScience* 27:1228.

Benton, J.M. and A.H. Cobb (1995). The plant growth regulator activity of the fungicide, epoxiconazole, on *Galium aparine* L. (cleavers). *Plant Growth Regulation* 17:149-155.

Benton, J.M. and A.H. Cobb (1997). The modification of phytosterol profiles and in vitro photosynthetic electron transport of *Galium aparine* L. (cleavers) treated with the fungicide epoxiconazole. *Plant Growth Regulation* 22:93-100.

Brown, R.G.S., H. Kawaide, and Y.Y. Yang (1997). Daminozide and prohexadione have similar modes of action as inhibitors of the late stages of gibberellin metabolism. *Physiologia Plantarum* 101:309-313.

Browning, G., A. Kuden, and P. Blake (1992). Site of (2RS,3RS)-paclobutrazol promotion of axillary flower initiation in pear cv Doyenne du Comice. *Journal for Horticultural Science* 67:121-128.

Browning, G., Z. Singh, A. Kuden, and P. Blake (1992). Effect of (2RS,3RS)-paclobutrazol on endogenous indole-3-acetic acid in shoot apices of pear cv Doyenne du Comice. *Journal for Horticultural Science* 67:129-135.

Burch, P.L., R.H. Wells, and W.N. Kline (1996). Red maple and silver maple growth evaluated 10 years after application of paclobutrazol tree growth regulator. *Journal of Arboriculture* 22:61-66.

Burrows, G.E., T.S. Boag, and W.P. Stewart (1992). Changes in leaf stem and root anatomy of chrysanthemum cv Lillian Hoek following paclobutrazol application. *Journal of Plant Growth Regulation* 11:189-194.

Butler, D.R., E. Pears, R.D. Child, and P. Brain (1989). Effects of triazole growth retardants on oilseed rape; photosynthesis of single leaves. *Annals of Applied Biology* 114:331-337.

Carter, J., B.P. Singh, and W. Whitehead (1996). Dikegulac but not benzyladenine enhances the aesthetic quality of Boston fern. *HortScience* 31:978-980.

Cathey, H.M. (1964). Physiology of growth retarding chemicals. *Annual Review of Plant Physiology* 15:271-302.

Cathey, H.M. (1975). Comparative plant growth retarding activities of ancymidol with ACPC, phosphon, chlormequat, and SADH on ornamental plant species. *HortScience* 10:204-216.

Cerdon, C., A. Rahier, M. Taton, and Y. Sauvaire (1995). Effects of tetcyclacis on growth and on sterol and sapogenin content in fenugreek. *Journal of Plant Growth Regulation* 14:15-22.

Chase, A.R., L.S. Osborne, and J.M.F. Yuen (1987). Effects of growth regulator chlormequat chloride on severity of three bacterial diseases on 10 cultivars of *Hibiscus rosa-sinensis*. *Plant Disease* 71:186-187.

Chong, C., A.G. Kanakis, and B. Bible (1982). Influence of growth regulators on ionic thiocyanate content of cruciferous vegetable crops. *Journal of the American Society for Horticultural Science* 107:586-589.

Cole, J.C. and R.M. Frymire (1995). The effects of uniconazole on woody plants. *American Nurseryman* 182:106+.

Conover, C.A. and L.N. Satterthwaite (1996). Paclobutrazol optimizes leaf size, vine length and plant grade of golden pothos (*Epipremnum aureum*) on totems. *Journal of Environmental Horticulture* 14:44-46.

Corr, B.E. and R.E. Widmer (1991). Paclobutrazol, gibberellic acid, and rhizome size affect growth and flowering of zantedeschia. *HortScience* 26:133-135.

Cosgrove, D.J. and S.A. Sovonick-Dunford (1989). Mechanism of gibberellin-dependent stem elongation in peas. *Plant Physiology* 89:184-191.

Cox, D.A. (1991). Gibberellic acid reverses effects of excess paclobutrazol on geranium. *HortScience* 26:39-40.

Cox, D.A. and G.J. Keever (1988). Paclobutrazol inhibits growth of zinnia and geranium. *HortScience* 23:1029-1030.

Cox, D.A. and F.F. Wittington (1988). Effects of paclobutrazol on height and performance of aluminum plant in a simulated interior environment. *HortScience* 23:222.

Cramer, C.S. and M.P. Bridgen (1998). Growth regulator effects on plant height of potted mussaenda 'Queen Sirikit'. *HortScience* 33:78-81.

Curry, E.A. and A.N. Reed (1989). Transitory growth control of apple seedlings with less persistent triazole derivatives. *Journal of Plant Growth Regulation* 8:167-174.

Davis, T.D., N. Sankhla, R.H. Walser, and A. Upadhyaya (1985). Promotion of adventitious root formation on cuttings by paclobutrazol. *HortScience* 20:883-884.

Davis, T.D., G.L. Steffens, and N. Sankhla (1988). Triazole plant growth regulators. *Horticultural Reviews* 10:63-105.

DeJong, T.M. and J.F. Doyle (1984). Leaf gas exchange and growth response of mature Fantasia nectarine trees to paclobutrazol. *Journal of the American Society for Horticultural Science* 109:878-882.

Deneke, C.F. and G.J. Keever (1992). Comparison of application methods of paclobutrazol for height control of potted tulips. *HortScience* 27:1329.

Deneke, C.F., G.J. Keever, and J.A. McGuire (1992). Growth and flowering of Alice du Pont mandevilla in response to sumagic. *Journal of Environmental Horticulture* 10:36-39.

Deneke, C.F., P.F. Thomas, and G.J. Keever (1992). Uniconazole restricts growth of seed-propagated *Physostegia virginiana* L. Benth 'Alba'. *HortScience* 27:928.

Dicks, J.W. (1972). Uptake and distribution of the growth retardant, daminozide, in relation to control of lateral shoot elongation in *Chrysanthemum morifolium*. *Annals of Applied Biology* 72:313-326.

Dicks, J.W. and D.A. Charles-Edwards (1973). A quantitative description of inhibition of stem growth in vegetative lateral shoots of *Chrysanthemum morifolium* by N-dimethylaminosuccinamic acid (daminozide). *Planta* 112:71-82.

Domir, S.C. and B.R. Roberts (1981). Trunk injection of plant growth regulators to control tree regrowth. *Journal of Arboriculture* 7:141-144.

Domir, S.C. and B.R. Roberts (1983). Tree growth retardation by injection of chemicals. *Journal of Arboriculture* 9:217-224.

Early, J.D. and G.C. Martin (1988). Sensitivity of peach seedling vegetative growth to paclobutrazol. *Journal of the American Society for Horticultural Science* 113:23-27.

Ecker, R., A. Barzilay, L. Aflin, and A.A. Watad (1992). Growth and flowering responses of *Matthiola incana* L. to paclobutrazol. *HortScience* 27:1330.

Einert, A.E. (1976). Slow-release ancymidol for poinsettia by impregnation of clay pots. *HortScience* 11:374-375.

El Hodairi, M.H., A.E. Canham, and W.R. Buckley (1988). The effects of paclobutrazol on growth and the movement of ^{14}C-labeled assimilates in Red delicious apple seedlings. *Journal for Horticultural Science* 63:575-581.

El-Keltawi, N.E. and R. Croteau (1986). Influence of ethephon and daminozide on growth and essential oil content of peppermint and sage. *Phytochemistry* 25:1285-1288.

El-Khoreiby, A.M., L.J. Lehman, and C.R. Unrath (1989). Adjuvant addition increases paclobutrazol spray efficacy on 'Delicious' apple. *HortScience* 24:1037.

Fisher, P.R., R.D. Heins, and J.H. Lieth (1996). Modeling the stem elongation response of poinsettia to chlormequat. *Journal of the American Society for Horticultural Science* 121:861-868.

Fletcher, R.A. and G. Hofstra (1985). Triadimefon: A plant multiprotectant. *Plant Cell Physiology* 26:775-780.

Fletcher, R.A. and G. Hofstra (1990). Improvement of uniconazole induced protection in wheat seedlings. *Journal of Plant Growth Regulation* 9:207-212.

Flint, H.M., J.E. Leggett, L. Elhoff, N.J. Parks, and E.W. Davidson (1996). Effects of the plant growth regulator mepiquat chloride on silverleaf whitefly (*Homoptera aleyrodidae*) infestation on cotton. *Journal of Entomological Science* 31:112-122.

Flores-Nimedez, A.A., K. Dorffling, and B.S. Vergara (1992). Improvement of chilling resistance in rice by application of an abscisic acid analog in combination with the growth retardant tetcyclacis. *Journal of Plant Growth Regulation* 12:27-34.

Foley, J.T. and G.J. Keever (1991). Growth regulators and pruning alter growth and axillary shoot development of Dianthus. *Journal of Environmental Horticulture* 9:191-195.

Foley, J.T. and G.J. Keever (1992). Pink polka-dot plant (*Hypoestes phyllostachya*) response to growth retardants. *Journal of Environmental Horticulture* 10:87-90.

Frymire, R.M. and J.C. Cole (1992). Uniconazole effect of growth and chlorophyll content of pyracantha, photinia, and dwarf Burford holly. *Journal of Plant Growth Regulation* 11:143-148.

Frymire, R.M. and J.C. Henderson-Cole (1992). Effect of uniconazole and limited water on growth water relations and mineral nutrition of *Landei pyracantha*. *Journal of Plant Growth Regulation* 11:227-231.

Fuller, K.P. and J. Zajicek (1995). Water relations and growth of vinca following chemical growth regulation. *Journal of Environmental Horticulture* 13:19-21.

Furukawa, K., Y.Y. Yang, I. Honda, T. Yanagisawa, A. Sakurai, N. Takahashi, and Y. Kamiya (1997). Effects of ethylene and gibberellins on the elongation of rice seedlings (*Oryza sativa* L). *Bioscience Biotechnology & Biochemistry* 61:864-869.

Furutani, S.C., E. Johnston, and M. Nagao (1989). Anthesis and abscission of blue jade vine flowers treated with ethephon and AOA. *HortScience* 24:1042.

Garber, M.P., W.G. Hudson, J.G. Norcini, R.K. Jones, A.R. Chase, and K. Bondari (1996). Pest management in the United States greenhouse and nursery industry: I. Trends in chemical and nonchemical control. *HortTechnology* 6:192-198.

Geneve, R.L. (1990). Root formation in cuttings of English ivy treated with paclobutrazol or uniconazole. *HortScience* 25:709.

Gent, M.P.N. (1995). Paclobutrazol or uniconazole applied early in the previous season promote flowering of field grown rhododendron and kalmia. *Journal of Plant Growth Regulation* 14:205-210.

Gent, M.P.N. (1997). Persistence of triazole growth retardants on stem elongation of rhododendron and kalmia. *Journal of Plant Growth Regulation* 16:197-203.

Gianfagna, T. (1995). Natural and synthetic growth regulators and their use in horticultural and agronomic crops. In *Plant Hormones*, P.J. Davies (Ed.). Netherlands: Kluwer Academic Publishers, pp. 751-773.

Gianfagna, T.J. and G.J. Wulster (1986). Comparative effects of ancymidol and paclobutrazol on Easter lily. *HortScience* 21:463-464.

Gilbertz, D.A. (1992). Chrysanthemum response to timing of paclobutrazol and uniconazole sprays. *HortScience* 27:322-323.

Gilley, A. and R.A. Fletcher (1997). Relative efficacy of paclobutrazol, propiconazole and tetraconazole as stress protectants in wheat seedlings. *Plant Growth Regulation* 21:169-175.

Gilliam, C.H., D.C. Fare, and J.T. Eason (1988). Control of *Acer rubrum* growth with flurprimidol. *Journal of Arboriculture* 14:99-101.

Grossmann, K. (1992). Plant growth retardants: Their mode of action and benefit for physiological research. *Current Plant Science* 13:788-797.

Grossmann, K., F. Siefert, J. Kwaikowski, M. Schraudner, C. Langebartels, and H. Sandermann (1993). Inhibition of ethylene production in sunflower cell suspensions by the plant growth retardant BAS111. Possible relations to changes in polyamine cytokinin contents. *Journal of Plant Growth Regulation* 12:5-11.

Gurusinghe, S.H. and K.A. Shackel (1995). Effect of ethephon (2-chloroethyl phosphonic acid) on vascular cambial strength of almond tree trunks. *Journal of the American Society for Horticultural Science* 120:194-198.

Hagiladi, A. and A.A. Watad (1992). *Cordyline terminalis* plants respond to foliar sprays and medium drenches of paclobutrazol. *HortScience* 27:128-130.

Hamada, M., T. Hosoki, and T. Maeda (1990). Shoot length control of tree peony (*Paeonia suffruticosa*) with uniconazole and paclobutrazol. *HortScience* 25:198-200.

Hamid, M.M. and R.R. Williams (1997). Translocation of paclobutrazol and gibberellic acid in Sturts desert pea (*Swainsona formosa*). *Plant Growth Regulation* 23:167-171.

Hartmann, H.T., D.E. Koster, F.T. Davies, and R.L. Geneve (Eds.) (1997). *Plant Propagation: Principles and Practices*, Sixth Edition. Englewood Cliffs, NJ: Prentice-Hall, pp. 288-289.

Harty, A.R. (1988). The use of growth retardants in citriculture. *Israel Journal of Botany* 37:155-164.

Haughan, P.A., R.S. Burden, J.R. Lenton, and L.J. Goad (1989). Inhibition of celery cell growth and sterol biosynthesis by the enantiomers of paclobutrazol. *Phytochemistry* 28:781-878.

Hedden, P. and Y. Kamiya (1997). Gibberellin biosynthesis enzymes, genes and their regulation. *Annual Review of Plant Physiology* 48:431-460.

Henderson, J.C. and T.H. Nichols (1991). Pyracantha coccinea 'Kasan' and 'Lalandei' response to uniconazole and chlormequat chloride. *HortScience* 26:877-880.

Hicklenton, P.R. (1990). Height control of pot chrysanthemums with pre- and post-plant treatments of daminozide and uniconazole. *Canadian Journal of Plant Science* 70:925-930.

Hield, H. (1979). Trunk bark banding with chlorfurenol for growth control. *Journal of Arboriculture* 5:59-61.

Ho, Y.S., K.C. Sanderson, and J.C. Williams (1985). Effect of chemicals and photoperiod on the growth and flowering of Thanksgiving cactus. *Journal of the American Society for Horticultural Science* 110:658-662.

Hofstra, G., L.C. Krieg, and R.A. Fletcher (1989). Uniconazole reduces ethylene and 1-aminocyclopropane-1-carboxylic acid and increases spermine levels in mung bean seedlings. *Journal of Plant Growth Regulation* 8:45-51.

Holcomb, E.J., S. Ream, and J. Reed (1983). The effect of BAS106, ancymidol and chlormequat on chrysanthemum and poinsettia. *HortScience* 18:364-365.

Holcomb, E.J., L.D. Tukey, and M.A. Rose (1991). Effect of GA on inflorescence in uniconazole-treated chrysanthemums. *HortScience* 26:312.

Holcomb, E.J. and J.W. White (1970). A technique for soil application of growth retardant. *HortScience* 5:16-17.

Horrell, B.A., P.E. Jameson, and P. Bannister (1990). Responses of ivy (*Hedera helix* L.) to combination of gibberellic acid, paclobutrazol and abscisic acid. *Plant Growth Regulation* 9:107-117.

Hunter, D.M. and J.T.A. Proctor (1994). Paclobutrazol reduces photosynthetic carbon dioxide uptake in grapevines. *Journal of the American Society for Horticultural Science* 119:486-491.

Izumi, K., Y. Kamiya, A. Sakurai, H. Oshio, and N. Takahashi (1985). Site of action of a uniconazole and comparative effects of its stereo isomers in a cell free system from cucurbit. *Plant and Cell Physiology* 26:821-827.

Jackson, M.J., M.A. Line, and O. Hasan (1996). Microbial degradation of a recalcitrant plant growth retardant—Paclobutrazol (PP333). *Soil Biology & Biochemistry* 28:1265-1267.

Jayroe-Counoyer, L. and S.E. Newman (1995). Stimulation of basal and axillary bud formation of container grown hybrid tea roses. *Journal of Environmental Horticulture* 13:47-50.

Jiao, J., X. Wang, and M.J. Tsujita (1990). Comparative effects of uniconazole drench and spray on shoot elongation of hybrid lilies. *HortScience* 25:1244-1246.

Johnson, C.R. (1974). Response of chrysanthemums grown in clay and plastic pots to soil application of ancymidol. *HortScience* 9:58.

Joiner, J.N., R.T. Poole, C.R. Johnson, and C. Ramcharam (1978). Effects of ancymidol and N, P, K on growth and appearance of *Dieffenbachia maculata* Baraquiniana. *HortScience* 13:182-184.

Kawabata, O. and R.A. Criley (1996). Dikegulac-sodium spray enhances uniform regrowth of *Murraya paniculata* L. Jack Hedge. *HortScience* 31:244-246.

Kawabata, O. and J. DeFrank (1994). A flexible function for regressing asymptotically declining responses of plant growth to growth retardant. *HortScience* 29:1357-1359.

Keever, G.J. (1990). Response of two forcing azalea cultivars to Bonzi and B-nine applications. *Journal of Environmental Horticulture* 8:182-184.

Keever, G.J. and D.A. Cox (1989). Growth inhibition in marigold following drench and foliar-applied paclobutrazol. *HortScience* 24:390.

Keever, G.J. and W.J. Foster (1989). Response of two florist azalea cultivars to foliar application of a growth regulator. *Journal of Environmental Horticulture* 7:56-59.

Keever, G.J. and W.J. Foster (1991a). Production and postproduction performance of uniconazole-treated bedding plants. *Journal of Environmental Horticulture* 9:203-206.

Keever, G.J. and W.J. Foster (1991b). Uniconazole suppresses bypass shoot development and alters flowering of two forcing azalea cultivars. *HortScience* 26:875-877.

Keever, G.J., W.J. Foster, and J.C. Stephenson (1990). Paclobutrazol inhibits growth of woody landscape plants. *Journal of Environmental Horticulture* 8:41-47.

Keever, G.J., C.H. Gilliam, and D.J. Eakes (1994). Cutless controls shoot growth of 'China Girl' holly. *Journal of Environmental Horticulture* 12:167-169.

Keever, G.J. and J.A. McGuire (1991). Sumagic (uniconazole) enhances flowering of Shishi-gashira camellia. *Journal of Environmental Horticulture* 9:185-187.

Keever, G.J. and J.W. Olive (1994). Response of Prize azalea to sumagic applied at several stages of shoot development. *Journal of Environmental Horticulture* 12:12-15.

Khalil, I.A. (1995). Chlorophyll and carotenoid contents in cereals as affected by growth retardants of the triazole series. *Cereal Research Communications* 23:183-189.

Kim, H.Y., T. Abe, H. Watanabe, and Y. Suzuki (1989). Changes in flower bud development of *Zinnia elegans* Jacq as influenced by the growth retardant S-07. *Journal for Horticultural Science* 64:81-89.

Kim, H.Y. and Y. Suzuki (1989). Changes in assimilated ^{14}C distribution and soluble acid invertase activity of *Zinnia elegans* induced by uniconazole, an inhibitor of gibberellin biosynthesis. *Plant Physiology* 90:316-321.

Kimball, S.L. (1990). The physiology of tree growth regulators. *Journal of Arboriculture* 16:39-41.

Konjevic, R., D. Grubisic, and M. Neskovic (1989). Growth retardant-induced changes in phototropic reaction of *Vigna radiata* seedlings. *Plant Physiology* 89:1085-1087.

Kraus, T.E., B.D. McKersie, and R.A. Fletcher (1995). Paclobutrazol-induced tolerance of wheat leaves to paraquat may involve increased antioxidant enzyme activity. *Journal of Plant Physiology* 145:570-576.

Kraus, T.E., D.P. Murr, and R.A. Fletcher (1991). Uniconazole inhibits stress induced ethylene in wheat soybean seedlings. *Journal of Plant Growth Regulation* 10:229-234.

Laiche, A.J. (1988). Effects of rate and repeat application of flurprimidol on the growth of *Photinia* x*fraseri* and *Ilex crenata* compacta. *Journal of Environmental Horticulture* 6:124-129.

Larrigaudiere, C., E. Pinto, and M. Vendrell (1996). Differential effects of ethephon and seniphos on color development of 'Starking Delicious' apple. *Journal of the American Society for Horticultural Science* 121:746-750.

Larson, R.A., C.B. Thorne, R.R. Milks, Y.M. Isenberg, and L.D. Brisson (1987). Use of ancymidol bulb dips to control stem elongation of Easter lilies grown in a pine bark medium. *Journal of the American Society for Horticultural Science* 112:773-777.

Latimer, J.C. and S.A. Baden (1994). Persistent effects of plant growth regulators on landscape performance of seed geraniums. *Journal of Environmental Horticulture* 12:150-154.

Latimer, J.C. and R.D. Oetting (1994). Paclobutrazol reduces insecticide phytotoxicity on salvia. *HortScience* 29:289-292.

Latimer, J.C., R.D. Oetting, and P.A. Thomas (1995). Method of application affects response of hollyhock to paclobutrazol. *HortScience* 30:626.

LeCain, D.R., K.A. Schekel, and R.L. Wample (1986). Growth-retarding effects of paclobutrazol on weeping fig. *HortScience* 21:1150-1152.

Lenton, J.R., N.E.J. Appleford, and K.E. Templesmith (1994). Growth retardant activity of paclobutrazol enantiomers in wheat seedlings. *Plant Growth Regulation* 15:281-291.

Lurie, S., R. Ronen, and B. Aloni (1995). Growth-regulator-induced alleviation of chilling injury in green and red bell pepper fruit during storage. *HortScience* 30:558-559.

Malek, A.A., F.A. Blazich, S.L. Warren, and J.E. Shelton (1992). Growth response of seedlings of flame azalea to manual and chemical pinching. *Journal of Environmental Horticulture* 10:28-31.

Martin, C.A., W.P. Sharp, J.M. Ruter, and R.L. Garcia (1994). Alterations in leaf morphology of two landscape shrubs in response to disparate climate and paclobutrazol. *HortScience* 29:1321-1325.

McAvoy, R.J. (1990). Poinsettia growth in media containing XE-1019 impregnated rockwool. *Acta Horticulturae* 272:215-222.

McAvoy, R.J. (1991). Response of Easter lily to preplant incorporation of uniconazole into the planting medium. *HortScience* 26:152-154.

McAvoy, R.J. and P. Kishbaugh-Schmidt (1992). Easter lily response to preplant incorporated ancymidol, in bark based media. *HortScience* 27:633.

McDaniel, G.L. (1983). Growth retardation activity of paclobutrazol on chrysanthemum. *HortScience* 18:199-200.

McDaniel, G.L. (1985).Transpiration in hydrangea as affected by antitranspirants and chlormequat. *HortScience* 20:293-296.

McDaniel, G.L. (1986). Comparison of paclobutrazol, flurprimidol, and tetcyclacis for controlling poinsettia height. *HortScience* 21:1161-1163.

McDaniel, G.L. (1990). Postharvest height suppression of potted tulips with paclobutrazol. *HortScience* 25:212-214.

McDaniel, G.L., E.T. Graham, and K.R. Maleug (1990). Alteration of poinsettia stem anatomy by growth-retarding chemicals. *HortScience* 25:433-435.

McKinless, J. and P.G. Alderson (1991). An anatomical study of rhizome bud formation induced by paclobutrazol and adventitious root formation in in vitro cultures of *Lapageria rosea* (Ruiz et Pav). *Annals of Botany* 67:331-338.

Messinger, N.L. and E.J. Holcomb (1986). The effect of chlormequat chloride, ancymidol, BAS106, and SD8339 on selected dianthus cultivars. *HortScience* 21:1397-1400.

Min, X.J. and D.P. Bartholemew (1996). Effect of plant growth regulators on ethylene production 1-aminocyclopropane-1-carboxylic acid oxidase activity and initiation of inflorescence development of pineapple. *Journal of Plant Growth Regulation* 15:121-128.

Miranda, R.M. and W.H. Carlson (1980). Effect of timing and number of applications of chlormequat and ancymidol on the growth and flowering of seed geraniums. *Journal of the American Society of Horticultural Science* 105:273-277.

Murray, G.E., K.C. Sanderson, and J.C. Williams (1986). Application methods and rates of ancymidol on plant height and seed germination of bedding plants. *HortScience* 21:120-122.

Needham, D.C. and P.A. Hammer (1990). Control of *Salpiglossis sinuata* height with plant growth regulators. *HortScience* 25:441-443.

Nelson, L.S. (1996). Growth regulation of Eurasian watermilfoil and flurprimidol. *Journal of Plant Growth Regulation* 15:33-38.

Newman, S.E. and J.S. Tant (1995). Root-zone medium influences growth of poinsettias treated with paclobutrazol-impregnated spikes and drenches. *HortScience* 30:1403-1405.

Newman, S.E., S.B. Tenney, and M.W. Follett (1989). Use of uniconazole to control height of *Hibiscus rosa-sinensis*. *HortScience* 24:1041.

Norcini, J.G. (1991). Growth and water status of pruned and unpruned woody landscape plants treated with sumagic (uniconazole), cutless (flurprimidol) and atrimmec (dikegulac). *Journal of Environmental Horticulture* 9:231-235.

Norcini, J.G., J.H. Aldrich, and J.M. McDowell (1994). Flowering response of bougainvillea cultivars to dikegulac. *HortScience* 29:282-284.

Norcini, J.G. and G.W. Knox (1989a). Effect of pruning on the growth inhibiting activity of sumagic (uniconazole). *Journal of Environmental Horticulture* 8:199-204.

Norcini, J.G. and G.W. Knox (1989b). Response of *Ligustrum* × *ibolium*, *Photinia* × *frazeri* and *Pyracantha koidzumii* 'Wonderberry' to XE-1019 and pruning. *Journal of Environmental Horticulture* 7:126-128.

Norman, S.M., R.D. Bennett, and S.M. Poling (1986). Paclobutrazol inhibits abscisic acid biosynthesis in *Cercospora rosicola*. *Plant Physiology* 80:122-125.

Orvos, A.R. and R.E. Lyons (1989). Photoperiodic inhibition of stem elongation and flowering in *Rudbeckia hirta* 'Marmalade'. *Journal of the American Society for Horticultural Science* 114:219-222.

Osborne, L.S. and A.R. Chase (1990). Chlormequat chloride growth retardant reduces spider mite infestations of *Hibiscus rosa-sinensis*. *HortScience* 25:648-650.

Owings, A.D. and S.E. Newman (1993). Chemical modification of *Photinia* × *fraseri* plant size and lateral branching. *Journal of Environmental Horticulture* 11:1-5.

Parivar, F., J.E. Preece, and G.D. Coorts (1985). The effects of ancymidol concentrations and application methods on cultivars of Mid-Century hybrid lily. *Journal for Horticultural Science* 60:263-268.

Pharis, R.P. and R.W. King (1985). Gibberellins and reproductive development in seed plants. *Annual Review of Plant Physiology* 36:517-568.

Pinhero, R.G. and R.A. Fletcher (1994). Paclobutrazol and ancymidol protect corn seedlings from high and low temperature stress. *Plant Growth Regulation* 15:47-53.

Purohit, A. and J.B. Shanks (1984). Effects of oxathiin on morphological development, histological changes and auxin activity within the shoot apex of chrysanthemum. *Journal of the American Society for Horticultural Science* 109:7-10.

Rademacher, W. (1991). Inhibitors of gibberellin biosynthesis: Applications in agriculture and horticulture. In *Gibberellins*, N. Takahashi, B.O. Phinney, and J. Macmillan (Eds.). New York: Springer-Verlag, pp. 296-310.

Rademacher, W. (1992a). The mode of action of α-cylcyclohexanediones—A new type of growth retardant. *Current Plant Science* 13:571-577.

Rademacher, W. (1992b). Inhibition of gibberellin production in the fungi *Gibberella fujikuroi* and *Sphaceloma manihoticola* by plant growth retardants. *Plant Physiology* 100:635-639.

Ramina, A. and P. Tonutti (1985). The effect of paclobutrazol on strawberry growth and fruiting. *Journal for Horticultural Science* 60:501-506.

Ranney, T.G., R.E. Bir, J.L. Conner, and E.P. Whitman (1994). Use of paclobutrazol to regulate shoot growth and flower development of *Roseum elegans* rhododendron. *Journal of Environmental Horticulture* 12:174-178.

Read, P.E., V.L. Herman, and D.A. Heng (1974). Slow-release chlormequat: A new concept in plant growth regulators. *HortScience* 9:55-57.

Rebers, M., G. Romeijn, and E. Knegt (1994). Effects of exogenous gibberellins and paclobutrazol on floral stalk growth of tulip sprouts isolated from cooled and non-cooled tulip bulbs. *Physiologia Plantarum* 92:661-667.

Reddy, A.R., K.R. Reddy, and H.F. Hodges (1996). Mepiquat chloride (PIX)-induced changes in photosynthesis and growth of cotton. *Plant Growth Regulation* 20:179-183.

Reed, A.N., E.A. Curry, and M.W. Williams (1989). Translocation of triazole growth retardants in plant tissues. *Journal of the American Society for Horticultural Science* 114:893-898.

Richardson, P.J., A.D. Webster, and J.D. Quinlan (1986). The effect of paclobutrazol sprays with or without the addition of surfactants on the shoot growth yield and fruit quality of the apple cultivars Cox and Suntan. *Journal for Horticultural Science* 61:439-446.

Rieger, M. and G. Scalabrelli (1990). Paclobutrazol, root growth, hydraulic conductivity, and nutrient uptake of 'Nemaguard' peach. *HortScience* 25:95-98.

Ritchie, G.A., K.C. Short, and M.R. Davey (1991). In vitro acclimatization of chrysanthemum and sugar beet plantlets by treatment with paclobutrazol and exposure to reduced humidity. *Journal of Experimental Botany* 42:1557-1563.

Roberts, C.M., G.W. Eaton, and F.M. Seywerd (1990). Production of fuchsia and tibouchina standards using paclobutrazol or chlormequat. *HortScience* 25:1242-1243.

Ruter, J.M. (1992). Growth and flowering response of butterfly-bush to paclobutrazol formulation and rate of application. *HortScience* 27:929.

Ruter, J.M. (1994). Growth and landscape establishment of pyracantha and juniperus after application of paclobutrazol. *HortScience* 29:1318-1320.

Ruter, J.M. (1996). Paclobutrazol application method influences growth and flowering of "New Gold" lantana. *HortTechnology* 6:19-20.

Sachs, R.M., M. Campidonica, J. Steffen, D. Hodel, and M.P. Jauniaux (1986). Chemical control of tree growth by bark painting. *Journal of Arboriculture* 12:284-291.

Sachs, R.M., J. DeBie, T. Kretchun, and T. Mock (1975). Comparative activity of commercially available maleic hydrazide formulations on several plant species. *HortScience* 10:366.

Sachs, R.M. and W.P. Hackett (1972). Chemical inhibition of plant height. *HortScience* 7:440-447.

Sachs, R.M., H. Hield, and J. DeBie (1975). Dikegulac: A promising new foliar-applied growth regulator for woody species. *HortScience* 10:367-369.

Sanderson, K.C. (1990). New application methods for growth retardants to media for production of clerodendrum. *HortScience* 25:125.

Sanderson, K.C., W.C. Martin, and J. McGuire (1988). Comparison of paclobutrazol tablets, drenches, gels, capsules, and sprays on chrysanthemum growth. *HortScience* 23:1008-1009.

Sanderson, K.J., W.C Martin, and J. McGuire (1990). New application methods for growth retardants to media for production of clerendron. *HortScience* 25:125.

Sanderson, K.C., W.C. Martin, and R.B. Reed (1989). Slow-release growth retardant tablets for potted plants. *HortScience* 24:960-962.

Sanderson, K.C., D.A. Smith, and J.A. McGuire (1994). Vacuum infusion with daminozide for retarding potted chrysanthemum height. *HortScience* 29:330.

Schneider, E.F. (1967). Conversion of the plant growth retardant (2-chloroethyl) trimethyl ammonium chloride to choline in shoots of chrysanthemum and barley. *Canadian Journal of Biochemistry* 45:395-400.

Schuch, U.K. (1994). Response of chrysanthemum to uniconazole and daminozide applied as a dip to cuttings or as a foliar spray. *Journal of Plant Growth Regulation* 13:115-121.

Schwartz, M.A., R.N. Payne, and G. Sites (1985). Residual effect of chlormequat on garden performance in sun and shade of seed and cutting-propagated cultivars of geraniums. *HortScience* 20:368-370.

Scnurr, J.P., Z.M. Cheng, and A. Boe (1996). Effects of growth regulators on sturdiness of Jack pine seedlings. *Journal of Environmental Horticulture* 14:228-230.

Shaw, P.M., K.A. Schekel, and V.I. Lohr (1991). Height control in pot-grown 'Wood Violet' gladiolus using ancymidol. *HortScience* 26:1089.

Shravan, D., M.R. Evans, and B.E. Whipker (1998). Paclobutrazol drenches control growth of potted sunflowers. *HortTechnology* 8:235-237.

Shukla, A., A.H. Abad Farooqi, Y.N. Shukla, and S. Sharma (1992). Effect of triacontanol and chlormequat on growth plant hormone and artemisinin yield of *Artemesia annua* L. *Plant Growth Regulation* 11:165-171.

Siefert, F., M. Thalmair, C. Lagebartels, H. Sandermann, and K. Grossmann (1996). Epiconazole induced stimulation of the antifungal hydrolases, chitinase and ß-1,3-glucanase in wheat. *Plant Growth Regulation* 20:279-286.

Singh, V.P. (1993). Uniconazole (S-3307) induced cadmium tolerance in wheat. *Journal of Plant Growth Regulation* 12:1-3.

Stamps, R.H. and R.J. Henny (1986). Paclobutrazol and night interruption lighting affect episcia growth and flowering. *HortScience* 21:1005-1006.

Stanley, C.J. and K.E. Cockshull (1989). The site of ethephon application and its effect on flower initiation and growth of chrysanthemum. *Journal for Horticultural Science* 64:341-350.

Starman, T.W. (1990). Whole-plant response of chrysanthemum to uniconazole foliar sprays or medium drenches. *HortScience* 25:935-937.

Starman, T.W. (1991). Lisianthus growth and flowering responses to uniconazole. *HortScience* 26:150-152.

Starman, T.W. (1993). Ornamental pepper growth and fruiting response to uniconazole depends on application time. *HortScience* 28:917-919.

Starman, T.W., T.A. Cerny, and T.L. Grindstaff (1994). Seed geranium growth and flowering responses to uniconazole. *HortScience* 29:865-867.

Starman, T.W. and P.T. Gibson (1992). Efficacy and postharvest persistence of uniconazole treatment on *Hypoestes phyllostachya*. *HortScience* 27:819-820.

Starman, T.W., J.W. Kelly, and H.B. Pemberton (1989). Characterization of ancymidol effects on growth and pigments of *Helianthus annuus* cultivars. *Journal of the American Society for Horticultural Science* 114:427-430.

Steinberg, S.L., J.M. Zajicek, and M.J. McFarland (1991a). Short-term effects of uniconazole on the water relations and growth of ligustrum. *Journal of the American Society for Horticultural Science* 116:460-464.

Steinberg, S.L., J.M. Zajicek, and M.J. McFarland (1991b). Water relations of hibiscus following pruning or chemical growth regulation. *Journal of the American Society for Horticultural Science* 116:465-470.

Sterrett, J.P. (1985). Paclobutrazol: A promising growth inhibitor for injection into woody plants. *Journal of the American Society for Horticultural Science* 110:4-8.

Sterrett, J.P. (1988). XE-1019: Plant response, translocation and metabolism. *Journal of Plant Growth Regulation* 7:19-26.

Sterrett, J.P. and T.J. Tworkoski (1987). Flurprimidol: Plant response translocation and metabolism. *Journal of the American Society for Horticultural Science* 112:341-345.

Sterrett, J.P., T.J. Tworkoski, and P.T. Kujawski (1989). Physiological responses of deciduous tree root collar drenched with flurprimidol. *Journal of Arboriculture* 15:120-124.

Stuart, N.W. (1961). Initiation of flower buds in rhododendron after application of growth retardants. *Science* 134:50-52.

Sugavanam, B. (1984). Diastereoisomers and enantiomers of paclobutrazol: Their preparation and biological activity. *Pesticide Science* 15:296-302.

Tagliavini, M. and N.E. Looney (1991). Response of peach seedlings to root zone temperature and root applied growth regulators. *HortScience* 26:870-872.

Takahashi, N., B.O. Phinney, and J. Macmillan (1991). *Gibberellins.* New York: Springer-Verlag, 426 pp.

Tayama, H.K. and S.A. Carver (1990). Zonal geranium growth and flowering responses to six growth regulators. *HortScience* 25:82-83.

Tayama, H.K. and S.A. Carver (1992a). Concentration response of zonal geranium and potted chrysanthemum to uniconazole. *HortScience* 27:126-128.

Tayama, H.K. and S.A. Carver (1992b). Residual efficacy of uniconazole and daminozide on potted 'Bright Golden Anne' chrysanthemum. *HortScience* 27:124-125.

Tezuka, T., C. Takahara, and Y. Yamamoto (1989). Aspects regarding the action of CCC in hollyhock plants. *Journal of Experimental Botany* 40:689-692.

Thetford, M., S.L. Warren, F.A. Blazich, and J.F. Thomas (1995). Response of *Forsythia* ×*intermedia* 'Spectabilis' to uniconazole: II. Leaf and stem anatomy, chlorophyll, and photosynthesis. *Journal of the American Society for Horticultural Science* 120:983-988.

Tija, B. (1976). Comparison of soil applied growth regulators on height control of poinsettia. *HortScience* 11:373-374.

Tjia, B. and T.J. Sheehan (1986). Chemical height control of *Lisianthus russellianus*. *HortScience* 21:147-148.

Tija, B., L. Stolz, M.S. Sandhu, and J. Buxton (1976). Surface active agent to increase effectiveness of surface penetration of ancymidol on hydrangea and Easter lily. *HortScience* 11:371-372.

Tworkoski, T.J. and J.P. Sterrett (1992). Phytotoxic effects, regrowth, and ^{14}C-sucrose translocation in Canada thistle treated with mefluidide, flurprimidol and systemic herbicides. *Journal of Plant Growth Regulation* 11:105-111.

Upadhyaya, A., T.D. Davis, M.H. Larsen, R.H. Walser, and N. Sankhla (1990). Uniconazole induced thermotolerance in soybean seedling root tissue. *Physiologia Plantarum* 79:78-84.

Upadhyaya, A., T.D. Davis, and N. Sankhla (1986). Some biochemical changes associated with paclobutrazol induced adventitious root formation on bean hypocotyl cuttings. *Annals of Botany* 57:309-315.

Van, T.K. and T.D. Center (1994). Effect of paclobutrazol and water hyacinth weevil (*Neochetina eichhorniae*) on plant growth and leaf dynamics of water hyacinth (*Eichornia crassipes*). *Weed Science* 42(4):665-672.

Viagro-Wolff, A.L. and M.R. Warmund (1987). Suppression of growth and plant moisture stress of forsythia with flurprimidol and XE-1019. *HortScience* 22:884-885.

Vu, J.C.V. and G. Yelenosky (1992). Growth and photosynthesis of sweet orange plants treated with paclobutrazol. *Journal of Plant Growth Regulation* 11:85-89.

Wainwright, H. and H.L. Irwin (1987). The effects of paclobutrazol and pinching on antirrhinum flowering pot plants. *Journal for Horticultural Science* 62:401-404.

Wample, R.L. and E.B. Culver (1983). The influence of paclobutrazol, a new growth regulator, on sunflowers. *Journal of the American Society for Horticultural Science* 108:122-125.

Wang, Q., C.H.A. Little, T. Moritz, and P.C. Oden (1995). Effects of prohexadione on cambial and longitudinal growth and the levels of endogenous gibberellins A_1 A_3 A_4 and A_9 and indole-3-acetic acid in *Pinus sylvestris* shoots. *Journal of Plant Growth Regulation* 14:175-181.

Wang, S.Y. and M. Faust (1986). Effect of growth retardants on root formation and polyamine content of apple seedlings. *Journal of the American Society for Horticultural Science* 111:912-917.

Wang, S.Y., T. Sun, and M. Faust (1986). Translocation of paclobutrazol, a gibberellin biosynthesis inhibitor, in apple seedlings. *Plant Physiology* 82: 11-14.

Wang, Y.T. (1987). Influence of ancymidol on growth and interior quality of *Syngonium podophyllum* 'White Butterfly'. *HortScience* 22:959-960.

Wang, Y.T. (1991). Growth stage and site of application affect efficacy of uniconazole and GA_3 in hibiscus. *HortScience* 26: 148-150.

Wang, Y.T. and T.M. Blessington (1990). Growth of four tropical foliage species treated with paclobutrazol or uniconazole. *HortScience* 25:202-204.

Wang, Y.T. and J.R. Dunlap (1994). Effect of GA_{4+7} on growth and cellular change in uniconazole-treated hibiscus. *Journal of Plant Growth Regulation* 13:33-38.

Wang, Y.T. and L.L. Gregg (1989). Uniconazole affects vegetative growth, flowering and stem anatomy of hibiscus. *Journal of the American Society for Horticultural Science* 114:927-932.

Wang, Y.T. and L.L. Gregg (1991). Modification of hibiscus growth by treating unrooted cuttings and potted plants with uniconazole or paclobutrazol. *Journal of Plant Growth Regulation* 10:47-51.

Wang, Y.T. and L.L. Gregg (1994). Chemical regulators affect growth, post-production performance and propagation of golden pothos. *HortScience* 29:183-185.

Wang, Y.T., K.H. Hsiao, and L.L. Gregg (1992). Antitranspirant, water stress, and growth retardant regulate growth of golden pothos. *HortScience* 27:222-224.

Wang, Y.T. and T.Y. Hsu (1994). Flowering and growth of phalaenopsis orchids following growth retardant applications. *HortScience* 29:285-288.

Warren, S.L. (1990). Growth response of 13 container-grown landscape plants to uniconazole. *Journal of Environmental Horticulture* 8:151-153.

Warren, S.L., F.A. Blazich, and M. Thetford (1991). Whole-plant response of selected woody landscape species to uniconazole. *Journal of Environmental Horticulture* 9:163-167.

Werbouck, S.P.O. and P.C. Debergh (1996). Imidazole fungicides and paclobutrazol enhance cytokinin induced adventitious shoot proliferation in Araceae. *Journal of Plant Growth Regulation* 15:81-85.

Werbrouck, S.P.O., P. Redig, H.A. Van Onckelen, and P.C. Debergh (1996). Gibberellins play a role in the interaction between imidazole fungicides and cytokinins in Araceae. *Journal of Plant Growth Regulation* 15:87-93.

Whealy, C.A., T.A. Nell, and J.E. Barrett (1988). Plant growth regulator reduction of bypass shoot development in azalea. *HortScience* 23:166-167.

Wheeler, N.C. (1987). Effect of paclobutrazol on Douglas fir and loblolly pine. *Journal for Horticultural Science* 62:101-106.

Whipker, B.E., R.T. Eddy, and P.A. Hammer (1994a). Chemical growth retardant application to lisianthus. *HortScience* 29:1368.

Whipker, B.E., R.T. Eddy, and P.A. Hammer (1994b). Chemical growth retardant height control of ornamental kale. *HortScience* 29:329.

Whipker, B.E., R.T. Eddy, and P.A. Hammer (1995). Chemical growth retardant application to tuberous-rooted dahlias. *HortScience* 30:1007-1008.

Whipker, B.E., R.T. Eddy, F. Heraux, and P.A. Hammer (1995). Chemical growth retardants for height control of pot asters. *HortScience* 30:1309.

Whitaker, B.D. and C.Y. Wang (1987). Effect of paclobutrazol and chilling on leaf membrane lipids on cucumber seedlings. *Physiologia Plantarum* 70:404-411.

White, J.W. and I.J. Warrington (1984a). Effects of split-night temperatures, light, and chlormequat on growth and carbohydrate status of *Pelargonium ×hortorum. Journal of the American Society for Horticultural Science* 109:458-463.

White, J.W. and I.J. Warrington (1984b). Growth and development responses of Geranium to temperature, light integral, CO_2, and chlormequat. *Journal of the American Society for Horticultural Science* 109:728-735.

Wichard, M. (1997). Paclobutrazol is phloem mobile in castor oil plant (*Ricinus communis* L). *Journal of Plant Growth Regulation* 16:215-217.

Wijayanti, L., S. Fujioka, M. Kobayashi, and A. Sakurai (1996). Effect of uniconazole and gibberellin on the flowering of *Pharbitis nil. Bioscience Biotechnolgy & Biochemistry* 60:852-855.

Wilfret, G.J., B.K. Harbaugh, and T.A. Nell (1978). Height control of pixie poinsettia with a granular formulation of ancymidol. *HortScience* 13:701-702.

Wilkins, H.F., K. Grueber, W. Healy, and H.B. Pemberton (1986). Minimum fluorescent light requirements and ancymidol interactions on the growth of Easter lily. *Journal of the American Society for Horticultural Science* 111:384-387.

Wilkinson, R.I. and B. Hanger (1992). Paclobutrazol in hydroponic solution advances inflorescence development of hydrangea 'Merritt's Supreme'. *HortScience* 27:1195-1196.

Wilkinson, R.I. and D. Richards (1987). Effects of paclobutrazol on growth and flowering of *Bouvardia humboldtii*. *HortScience* 22:444-445.

Wilkinson, R.I. and D. Richards (1991). Influence of paclobutrazol on growth and flowering of rhododendron 'Sir Robert Peel'. *HortScience* 26:282-284.

Williamson, J.G., D.C. Coston, and L.W. Grimes (1986). Growth responses of peach roots and shoots to soil and foliar-applied paclobutrazol. *HortScience* 21:1001-1003.

Wood, B.W. (1984). Influence of paclobutrazol on selected growth and chemical characteristics of young pecan seedlings. *HortScience* 19:837-839.

Woolf, A.B., J. Clemens, and J.A. Plummer (1992). Selective removal of floral buds from camellia with ethephon. *HortScience* 27:32-34.

Woolf, A.B., J. Clemens, and J.A. Plummer (1995). Leaf maturity and temperature affect the selective removal of floral buds from camellia with ethephon. *Journal of the American Society for Horticultural Science* 120:614-621.

Wright, D.C. and J.T. Moran (1988). Adverse effects of Atrinal (dikegulac) on plane tree, red maple and Norway maple. *Journal of Arboriculture* 14:125-128.

Wulster, G.J. and T.J. Gianfagna (1991). *Freesia hybrida* response to ancymidol, cold storage of corms, and greenhouse temperatures. *HortScience* 26:1276-1278.

Wulster, G.J., T.J. Gianfagna, and B.B. Clarke (1987). Comparative effects of ancymidol, propiconazol, triadimefon, and Mobay RSW0411 on lily height. *HortScience* 22:601-602.

Yelenosky, G., J.C.V. Vu, and H.K. Wutscher (1995). Influence of paclobutrazol in the soil on growth, nutrient elements in the leaves, and flood freeze tolerance of citrus rootstock seedlings. *Journal of Plant Growth Regulation* 14:129-134.

Yim, K.O., Y.W. Kwon, and D.E. Bayer (1997). Growth responses and allocation of assimilates of rice seedlings by paclobutrazol and gibberellin treatment. *Journal of Plant Growth Regulation* 16:35-41.

Yokota, T., Y. Nakamura, N. Takahashi, M. Nonaka, H. Sekimoto, H. Oshio, and S. Takatsuto (1991). Inconsistency between growth and endogenous levels of gibberellins, brassinosteroids and sterols in *Pisum sativum* treated with uniconazole enantiomers. In *Gibberellins*, N. Takahashi, B.O. Phinney, and J. Macmillan (Eds.), New York: Springer-Verlag, pp. 339-349.

Zhang, S.A., J.T. Cothren, and E.J. Lorenz (1990). Mepiquat chloride seed treatment and germination temperature effects on cotton growth, nutrient partitioning and water use efficiency. *Journal of Plant Growth Regulation* 9:195-199.

Chapter 5

Role of Growth Regulators in the Postharvest Life of Ornamentals

Margrethe Serek
Michael S. Reid

Natural and artificial regulators of plant growth and development have been tested widely and used commercially to improve the postproduction quality of many perishable ornamental plants, including cut flowers and foliage, potted flowering plants, and potted foliage plants. Of particular and recent importance has been the use of inhibitors of ethylene synthesis for extending the life of flowers and potted plants that are sensitive to ethylene. In this chapter we discuss the factors affecting the life of ornamentals, and the way in which growth regulators have been used to extend it.

POSTHARVEST LIFE OF ORNAMENTALS

Plant materials that are grown and harvested for their ornamental value cover a very wide range of taxa, including ferns and fern allies, gymnosperms, and angiosperms. The products that we consider ornamentals include those which are cut for their flowers and/or foliage and those which are harvested intact as potted flowering plants or potted foliage plants. Dormant ornamental nursery plants and propagules, including bulbs, corms, tubers, and rhizomes, also are considered ornamental crops, but these are outside the scope of this chapter.

Whether cut or intact ornamentals, the loss of quality in stems, leaves, or flower parts may signal an end of display life. Premature wilting of stem tips or leaves in cut foliage, while very different in nature from the programmed wilting of ethylene-sensitive flowers, is equally cause for rejection in the marketplace. In some ornamentals, loss of quality may result from one of several causes, including wilting or abscission of leaves and/or petals, yellowing of leaves, and geotropic or phototropic bending of scapes and stems. In any consideration of the use of growth regulators to improve the postharvest life and quality of ornamentals, it is important to dissect these various aspects of quality loss and reduced postharvest life.

Growth, Development, and Aging

In plants, death of individual organs, and of the whole plant itself, is an integral part of the life cycle. Reproductive organs, particularly, have a defined program of growth, maturation, and senescence that is integrated with attraction of a pollinator, pollination, and maturation of the seed. Recent studies have shown that the events of flower senescence are programmed at the level of the genome (Woodson et al., 1993; Valpuesta et al., 1995). Even in the absence of senescence of floral organs or leaves, the continuing growth process can result in quality loss, for example, in foliage plants that lose their attractive form, or in flowering plants in which internode expansion gives an undesirable shape.

Flower Senescence

The senescence of flower petals or florets is a common cause of quality loss and reduced vase life for flowering plants and cut flowers alike. Flowers can be divided into several categories in terms of their senescence. Some flowers are extremely long-lived, especially in the Asteraceae and Orchidaceae. Others are short-lived, including many ornamental geophytes, such as *Tulipa, Iris,* and *Narcissus.* In a number of species, the flowers are ephemeral, lasting one day or less. Among the ephemerals are some important ornamentals, such as *Hibiscus* and *Hemerocallis* (the "day" lily).

The role of growth regulators in flower senescence has been the subject of considerable investigation. Clearly, senescence of many flowers is coordinated by a rise in ethylene biosynthesis. The life of such flowers can be improved through manipulation of their production and sensitivity to ethylene. In another group, the so-called nonclimacteric or ethylene-insensitive flowers, researchers have not yet elucidated the mechanisms associated with the induction of senescence. The different systems controlling senescence do not appear to be correlated with the longevity of the flower. Senescence in the ephemeral flowers of morning glory and hibiscus is coordinated by a rise in ethylene evolution, but so is the senescence of the very long-lived flowers of *Phalaenopsis*.

Water Loss and Wilting

Extended life for cut and potted ornamentals depends absolutely on a continuing and adequate supply of water. Failure of water supply, whether through obstruction of the cut stems or inadequate watering of pots, results in rapid wilting of shoot tips, leaves, and petals.

Leaf Yellowing and Senescence

Yellowing of leaves, and even of other organs (buds, pedicels, and stems), commonly is associated with the end of display life in some ornamentals. Leaf yellowing is a complex physiological process that may be a response to one of several inducers. Although ethylene exposure will often cause leaf yellowing, inhibition of ethylene action may not prevent it (see Table 5.1). Early research demonstrated

TABLE 5.1. Effect of Inhibitors of Ethylene Action on Leaf Yellowing of *Rosa hybrida* 'Royal' Grown in California or in Denmark

	Treatment	Control	1-MCP	STS
California roses	Air	17.4a*	15.9a	13.4a
	Darkness	31.9b	34.5b	38.4b
Danish roses	Air	10.5ab	8.4b	11.3ab
	Darkness	14.6a	11.8ab	14.7a

Source: Serek, Sisler, and Reid, 1996, p. 148.

Note: 1-MCP = 1-methylcyclopropene; STS = silver thiosulfate.

*Mean separation within columns and rows at $P = 0.05$, according to Student's t test for the hypothesis H_0:LSM(i) = LSM(j).

the ability of cytokinins to inhibit leaf yellowing, and a cytokinin treatment is occasionally prescribed to overcome this problem in cut chrysanthemums.

Abscission

Complex interactions among growth regulators also are involved in the induction of abscission, another physiological event that may result in quality loss in ornamentals. In miniature rose plants, for example, low light, exposure to ethylene, drought, or other stresses may result in catastrophic abscission of leaves, petals, and even buds. *Pelargonium* flowers are an extreme example of such sensitivity (see Photo 5.1).

FACTORS AFFECTING
THE POSTHARVEST LIFE OF ORNAMENTALS

Despite the diversity of taxa and even plant parts that make up what we term ornamental crops, a number of common factors affect many, or all, of them. As a preface to discussing the effects of growth regulators on the postharvest life of ornamentals, it is appropriate to review briefly the most important factors affecting their quality and ornamental life.

Flower Maturity

Since potted plants can continue to grow after transportation and sale, the stage at which they are taken from the greenhouse is primarily an aesthetic and marketing question. Typically, flowering plants will be harvested when some flowers are showing color. Foliage plants will be harvested when they have reached an aesthetically pleasing stage. Because cut flowers are separated from their support system, minimum harvest maturity can be defined as the stage at which buds on the harvested bloom can be opened fully and have satisfactory display life after distribution. Many flowers are presently harvested when the buds are starting to open (rose, gladiolus), although others are normally fully open or nearly so (chrysanthemum, carnation). Flowers for local markets are generally harvested more fully open than those intended for storage and/or long-distance transport.

Temperature

Plant respiration, which supplies the energy for growth and aging, generates heat as a by-product. Furthermore, as the ambient temperature rises, the respiration rate increases. For example, ornamentals kept at 30°C are likely to respire (and therefore age) up to forty-five times as fast as those kept at 0°C. The rate of aging can be dramatically reduced by cooling the flowers (see Photo 5.2). The optimum temperature for storage of common cut flowers is near the freezing point (0°C). Some tropical crops, such as anthurium, bird-of-paradise, some orchids, and ginger, however, are injured at temperatures below 10°C. Symptoms of this "chilling injury" include darkening of the petals, water soaking of the petals (which look transparent), and, in severe cases, collapse and drying of leaves and petals (see Photo 5.3). Most foliage plants, and some important flowering plants, are very sensitive to chilling temperatures. Foliage plants, in general, are selected from tropical jungles, where they are adapted to warm, humid conditions. They are usually destroyed by even brief exposure to temperatures below 10°C. Among the important flowering potted plants that are sensitive to low temperature are poinsettia, bromeliad, African violet, and anthurium.

Food Supply

Starch and sugar stored in the stem, leaves, and petals provide much of the food needed for flower opening and maintenance. These carbohydrate levels are highest when plants are grown in high light and with proper cultural management. Carbohydrate levels are, in fact, generally highest in the late afternoon—after a full day of sunlight. However, cut flowers and potted plants are preferably harvested in the early morning because temperatures are low, plant water content is high, and a whole day is available for processing the crop.

Light

The postharvest life of potted plants is limited by their ability to maintain photosynthesis under the lighting conditions of the interior environment where they are kept. If the light integral is below the light photosynthetic compensation point, the plant will soon die. Light levels are not as important for cut flowers, except in cases in which foliage yellowing is a problem. Leaves of some cut flowers, such as chrysanthemum, marguerite daisy, lily, and alstroemeria may yellow rapidly, especially if stored in darkness at warm temperatures. We have shown that the blackening of leaves of cut flowers of *Protea nerifolia* is induced by low carbohydrate status in the harvested inflorescence and can be prevented by maintaining the flowers in high light.

Water Supply

Typical ornamentals, especially those with leafy stems, have a large surface area, so they lose water and wilt very rapidly. They should be stored at relative humidities above 95 percent to minimize water loss, particularly during long-term storage. Water loss is dramatically reduced at low temperatures, another reason for prompt and efficient cooling of cut flowers and potted plants. Even after cut flowers or potted plants have lost considerable water (for example, during transportation or storage), they can be fully rehydrated using proper techniques. Cut flowers will absorb solutions without difficulty, providing nothing obstructs water flow in the stems. Even a severely desiccated potted plant can recover if provided with adequate water.

Ethylene

Many ornamentals are adversely affected by exposure to ethylene gas. In a survey of a wide variety of taxa, Woltering and Van Doorn (1988) distinguished different classes of response. Some ornamentals, such as members of the Asteraceae (*Dendranthema, Liatris, Aster*), are unaffected by ethylene. In others, as with many members of the Orchidaceae and Caryophyllacae, exogenous and endogenous ethylene is associated with accelerated wilting and changes in color of floral organs. In a wide range of taxa, ethylene is associated with accelerated abscission of petals, flowers, and/or inflorescences. In plants that respond to ethylene, the presence of this gas in the environment often causes reduced quality and shortened display life, as a result of these physiological responses. In foliage plants, too, ethylene can cause undesirable responses, including abscission of leaves (for example, in *Cissus*, which is extremely sensitive to this gas) and leaf yellowing or discoloration (as in *Asplenium nidum*).

Growth Tropisms

Certain responses of ornamentals to environmental stimuli (tropisms) can result in quality loss. Most important are geotropism (bending away from gravity) and phototropism (bending toward light). Geotropism often reduces quality in spike-type cut flower crops, such as gladiolus and snapdragon, because the flowers and spike bend upward when stored horizontally (see Photo 5.4). Phototropism and stretching of scapes and spikes can be caused by directional light or low light during marketing of flowering plants at warm temperatures.

Mechanical Damage

Bruising and breakage of ornamental crops should always be avoided. Flowers with torn petals, broken stems, or other obvious injuries are undesirable for aesthetic reasons. Disease organisms can more easily infect plants through injured areas. In fact, many disease organisms can only enter a plant through an injury point. In addition, respiration and ethylene evolution is generally higher in injured plants, further reducing storage and vase life.

Disease

Ornamentals are very susceptible to disease, not only because their petals are fragile, but also because dead or dying flower or foliage parts and nectary secretions provide an excellent nutrient supply for even mild pathogens. To make matters worse, transfer from cold storage to warmer handling areas can result in condensation of water on the harvested flowers. The most commonly encountered disease organism, gray mold (*Botrytis cinerea*), can germinate wherever free moisture is present. In the humid environment of the flower head, it can even grow (albeit more slowly) at temperatures near freezing. Proper management of greenhouse hygiene, temperature control, and the minimizing of condensation on harvested crops all reduce losses caused by this disease.

GROWTH REGULATORS
AND THE POSTHARVEST LIFE OF ORNAMENTALS

With this understanding of the factors affecting the postharvest quality and life of cut flowers and potted plants, we can examine the role and possible utility of the plant hormones and other plant growth regulators in these processes.

Auxin and Inhibitors of Auxin Transport

Bending of stems away from gravity, an important cause of quality loss in spike-type flowers such as snapdragon, gladiolus, and red-hot poker (*Kniphofia*) (Woltering et al., 1991), is the effect of redistribution of auxin in response to gravity. Negative geotropism in such flowers can be prevented by a pulse treatment with naphthyl pthalamic acid (NPA), an inhibitor of auxin transport. In our experiments (see Photo 5.5), this treatment proved very effective in red-hot poker and in snapdragons.

Auxin is also implicated in another tropic response, the downward bending of the petioles of poinsettia bracts in response to sleeving. Initially, researchers considered this response to be an ethylene response. Sleeving does stress the petioles, and this stress results in increased ethylene biosynthesis (Saltveit, Pharr, and Larson, 1979; Staby et al., 1981). It is well known that ethylene can cause epinastic curvature (see Photo 5.6), and these researchers concluded that the endogenous stress-induced ethylene caused this phenomenon in sleeved poinsettias. Attempting a practical solution to this problem, we treated poinsettia plants with silver thiosulfate (STS) (Reid, Mor, and Kofranek, 1981) and discovered that this potent inhibitor of ethylene action was without effect on poinsettia sleeving-induced epinasty. Further investigation revealed that changed distribution of auxin, not ethylene, caused the response. It would be worth testing NPA as a possible tool for overcoming this particular response.

Another subtle and commercially important effect of auxins is their role in the process of abscission. The literature on abscission abounds with exceptions and special cases, but the results of many of the studies on the effects of environmental conditions and applied plant hormones are consistent with the following hypothesis (Reid, 1985):

- A decreasing gradient in the auxin concentration between the subtended organ and the plant axis maintains the abscission zone in an insensitive state. The gradient is modulated by factors that modify the senescence of the organ. This is thought to be the reason that auxins, cytokinins, light, and good nutrition all tend to reduce or delay abscission.
- When the auxin is reduced or reversed, the abscission zone becomes sensitive to ethylene. When auxin is applied on the axial side of the abscission zone, when the leaf blade is removed, or if the plant is subjected to treatments that accelerate its senescence (shading, poor nutrition, ethylene), abscission is hastened. One of the effects of ethylene, or of stresses that enhance its production, in stimulating abscission may be by reducing auxin synthesis, and/or by interfering with its transport from the leaf, thus reducing the auxin gradient. In systems where abscisic acid (ABA) stimulates abscission, it may do so by interfering with the production, transport, or action of auxin, and/or by stimulating ethylene production.
- Once the cells of the abscission zone are sensitized, they respond to low concentrations of exogenous or endogenous ethylene by the rapid synthesis and secretion of hydrolytic enzymes. This activity results in the abscission of the target organ.

Abscission, a common factor for quality loss in cut flowers and potted plants, and usually attributed directly to ethylene, is thus primarily a response to changed auxin distribution patterns. Although inhibitors of ethylene action are prompt and effective tools for overcoming abscission, treatment of the plants to maintain the auxin gradient (freedom from stress, adequate light, absence of exogenous ethylene) is an important additional tool, especially in the production environment.

Natural and Artificial Cytokinins

The demonstration that benzyladenine could prevent dark-induced yellowing of leaves was a parlor trick that amused many plant physiologists in the 1950s. Included in numerous laboratory notebooks was a dried tobacco leaf on which the investigator's name was signed in 6-benzyladenine (BA). In the 1960s, it finally was established that this compound was stimulating processes that were naturally regulated by the plant hormone zeatin. In addition to their important roles in controlling and stimulating cell division, zeatin and its analogs also inhibit leaf senescence. Some evidence (Zacarias and Reid, 1990) suggests that the presence of high levels of cytokinin may reduce the sensitivity of leaf tissue to ethylene.

Leaf yellowing is an important factor in loss of quality of potted plants and cut flowers. We have shown (Serek, Sisler, and Reid, 1996) that inhibition of ethylene action by STS and other inhibitors may not necessarily prevent leaf yellowing and senescence in potted rose plants. The simultaneous application of STS and 6-benzylaminopurine (Tjosvold, Wu, and Reid, 1994) markedly improved plant quality over application of either regulator alone (see Photo 5.7).

Leaf yellowing in cut chrysanthemums also has been ameliorated by the application of exogenous cytokinins as a spray or dip. Some potted chrysanthemum cultivars are prone to leaf yellowing, and here, too, cytokinins have successfully been used to reduce the problem. Steinitz and colleagues (1980) studied factors controlling the retardation of chlorophyll degradation during senescence of cut pieces of stems of statice (*Limonium sinuatum*). They showed that the accelerated yellowing following harvest was a phytochrome-mediated response, and that it could be reduced by immersing the stem sections in high concentrations (up to 1 millimolar [mM]) BA.

Application of growth retardants to potted plants frequently will increase their susceptibility to leaf yellowing. This is a common problem in the case of Easter lilies (*Lilium longiflorum*). Recently, "Promalin," a growth-regulating chemical formulated with cytokinin and gibberellin components to provide "typier" apples, has been shown to be a very effective means of reducing leaf yellowing in lilies that have been treated with growth retardants (Han, 1997).

The utility of cytokinins for preventing leaf yellowing has suggested a molecular approach to overcoming postharvest leaf yellowing in plants. Researchers at the University of Wisconsin have isolated senescence-associated genes (SAGs) from *Arabidopsis thaliana*. Some of these genes are strongly up-regulated in the early phases of leaf senescence. By transforming tobacco plants with a construct combining the SAG promoter with the structural gene encoding isopentenyl transferase (a key enzyme in the cytokinin biosynthesis pathway), these researchers have produced plants in which the onset of senescence is indefinitely postponed (Gan and Amasino, 1995, 1997).

Ethylene

Last century, it was observed that leaking gas pipes or heaters in floriculture greenhouses caused many symptoms of toxicity, including early wilting (sleepiness) of carnations and abscission of leaves, florets, and petals from many other taxa. Ethylene, the active principal in the illuminating gas, was not identified as a plant growth regulator until the seminal work of Neljubow (1901). The dramatic effects of this gas on flower senescence were intensively studied by many researchers in the 1930s, particularly by those at the Boyce Thompson Institute. In a series of publications, these researchers demonstrated the effects of ethylene (albeit at very high concentrations) on wilting of carnations, coloration of orchids, abscission of leaves, and other processes. Although it had already been shown in 1934 that ethylene was produced by ripening fruits (Gane, 1934), not until the advent of the gas chromatograph as a tool for measuring low concentrations of ethylene was its hormonal nature accepted (Goeschl, Rappaport, and Pratt, 1966). Nichols (1966; 1977) first demonstrated, in carnations, the role of ethylene in natural flower senescence. The rapid advances in knowledge of the biochemistry and molecular biology of ethylene biosynthesis and action in the last two decades have provided a range of tools for overcoming the many deleterious effects of this gas on ornamental plants.

Silver Ion

Beyer (1976a, b; 1978) first reported the impressive effects of silver ion in inhibiting the action of ethylene. His data included a demonstration of the retardation of ethylene-induced senescence of carnations and orchids (Beyer, 1976a). At that time, these effects were primarily of interest to plant physiologists, since the unsightly damage caused to petals by dipping them in silver salts made the treatment commercially infeasible. Attempts to provide silver (Ag) through the stem base were unsuccessful, due to the immobility of the silver ion.

Veen and Van de Geijn (1978) reported that the extremely stable anionic silver thiosulfate complex ($Kd = 10^{-17}$) *was* mobile in cut stems yet provided sufficient silver to petals to prevent the effects of ethylene. Horticulturists finally had a tool that provided a practical means of preventing the effects of exogenous ethylene, one that would extend the life of ornamentals whose natural senescence was coordinated by ethylene (Reid et al., 1980). Dramatic improvements in vase life were subsequently reported for many cut flowers, including snapdragon (*Antirrhinum majus*), gypsophila (*Gypsophila paniculata*), sweet pea (*Lathyrus odoratus*), lily (*Lilium* spp.), and many other species (Veen, 1983; Mor, Reid, and Kofranek, 1984; Reid and Evans, 1988; Borochov and Woodson, 1989; Reid, Evans, and Dodge, 1989; Joyce, Reid, and Evans, 1990; Reid and Wu, 1992; Van Doorn and Reid, 1992). Silver thiosulfate quickly became a standard commercial treatment for ethylene-sensitive flowers. Applications could be short- or long-term; the key was to provide approximately 1 micromole (μmol) Ag^+ per stem. Colombian producers quickly adopted this new treatment and sold their product as "super carnations." The beneficial effects of STS catapulted carnations from their status as one of the shorter-lived flowers to a florist staple that could be relied on to last at least ten days.

Soon after the demonstration of the beneficial effects of pulse treatments with STS on cut flowers, researchers tested this material with potted plants, some of which are particularly sensitive to ethylene. The dramatic effects of an STS spray on the quality of zygocactus (*Schlumbergera truncata*) plants (see Photo 5.8) was demonstrated by Cameron and Reid (1981), who also reported beneficial effects for a range of other flowering potted plants, including *Impatiens*, *Calceolaria*, and *Pelargonium* species. The beneficial effects of STS resulted in the commercialization of a range of products, including concentrates, diluters, two-component concentrates, and a very convenient powdered formulation (Agrylene).

Because STS contains silver, which is an environmental pollutant, some countries have restricted its use, principally in relation to disposal of waste treatment solutions. Researchers have therefore been seeking alternative tools to control the effects of ethylene on the display life of potted flowering plants and cut flowers.

Inhibitors of Ethylene Biosynthesis: AVG and AOA

The rate-limiting step in the biosynthetic pathway for ethylene is the synthesis of 1-aminocyclopropane carboxylic acid (ACC) from S-adenosyl methionine (Yang and Hoffman, 1984), a step that involves pyridoxal phosphate as a cofactor. This explained the inhibitory effects of a bacterial toxin, aminoethoxyvinyl glycine (AVG), which had already been reported by Lieberman and his co-workers, and which had been used by Wang and colleagues (1977) to extend the life of carnations and other ethylene-sensitive flowers. It also suggested the use of a cheaper chemical inhibitor of pyridoxal phosphate-requiring enzymes, aminooxyacetic acid (AOA). Fujino and colleagues (1981) demonstrated the benefits of this material in extending the life of carnations. Other possible inhibitors of ethylene biosynthesis that have been tested include inhibitors of the conversion of ACC to ethylene by ACC oxidase. Cobalt (Co) ion, which has been reported to extend the life of roses and other flowers, is known to inhibit ACC oxidase. It also is known to be biocidal, and the results with cut flowers are as likely to be due to improved water relations as they are to be related to inhibition of ethylene biosynthesis, particularly in roses, in which senescence is not normally considered to be an ethylene-mediated process. The isolation of a naturally occurring inhibitor of ACC oxidase (Shih, Dumbroff, and Thompson, 1989) could provide a tool to separate the biocidal and inhibitory effects of Co^{2+}.

A recent commercial replacement for STS treatment of cut flowers that was developed in Holland uses an inhibitor of ethylene biosynthesis (Harkema, Dekker, and Essers, 1991; Woltering et al., 1987) to prevent accumulation of endogenous ethylene. We tested the practical benefits of AVG and AOA, and our data showed that these inhibitors of ethylene biosynthesis were unable to improve the display life of plants and flowers exposed to external ethylene (Serek and Andersen, 1993; Serek and Reid, 1993; Staby et al., 1993). Reduction in endogenous production of ethylene may be useful during marketing, when plants are exposed to stresses that induce endogenous ethylene production. However, since plants are frequently also exposed to exogenous ethylene during marketing, pretreatment with inhibitors of ethylene action seems to be a better strategy.

1-MCP

We have been exploring possible substitutes for STS, particularly gaseous ethylene analogs synthesized by Sisler that appear to inactivate the ethylene binding site irreversibly (Sisler, 1977, 1991; Sisler, Dupille, and Serek, 1996; Sisler, Serek, and Dupille, 1996; Sisler and Serek, 1997). The possibility that ethylene analogs might be effective inhibitors of ethylene action was first suggested by the inhibitory effects of the cyclic di-olefine norbornadiene (Sisler, Reid, and Fujino, 1983; Sisler and Yang, 1984). This material was not commercially practicable, since inhibition required the continuous presence of the inhibitor. Diazocyclopentadiene (DACP), initially synthesized as a possible photo-affinity label for the ethylene binding site, proved to be a very effective and irreversible inhibitor of ethylene action (Sisler and Blankenship, 1993; Sisler and Lallu, 1994). We found this material to be effective in overcoming the action of ethylene in roses (see Photo 5.9) and in geranium (Serek, Sisler, and Reid, 1993, 1994a). This material had limited commercial potential because of the difficulties inherent in handling an unstable and potentially explosive chemical.

1-MCP (1-methylcyclopropene), a cyclic olefine that may be one of the photodecomposition products of DACP, appears to have considerable commercial potential as an inhibitor of ethylene action. Preliminary data indicate that the compound is nontoxic and quite stable under normal conditions. This material has proved to be an effective inhibitor of ethylene effects in ethylene-sensitive potted flowering plants and cut flowers (Serek, Sisler, and Reid, 1994b, 1995). Concentrations of 1-MCP as low as 6 parts per billion (ppb) can prevent the detrimental effects of subsequent exposure to ethylene concentrations as high as 1 part per million (ppm). It also inhibits natural ethylene-induced senescence, including, for example, the induction of senescence in orchid flowers by removal of the pollinia (see Photo 5.10).

Our studies with 1-MCP suggest that this compound inhibits ethylene action by binding irreversibly to the ethylene binding site. Pretreatment of cut carnations (*Dianthus caryophyllus* L.) with 3 nanoliters per liter (nl·liter⁻¹) 1-MCP for 6 hours (h) inhibited their normal wilting response when they were exposed continuously to 0.4 microliters per liter (µl·liter⁻¹) ethylene. The 1-MCP-treated carnations also lasted longer than control flowers when held in ethylene-free air, presumably because the action of endogenously produced ethylene was also inhibited. Potted begonia plants (*Begonia* ×*elatior hybrida* 'Rosa') and cut penstemon flowers (*Penstemon hartwegii* Benth. ×*P. cobaea* Nutt.) treated with 1-MCP were also protected from the effects of exogenous ethylene, which normally causes rapid abscission of buds and flowers. Treatment with 1-MCP was at least as effective as treatment with STS.

The dramatic inhibition of the deleterious effects of ethylene in potted plants and cut flowers by 1-MCP pretreatment indicates that this compound, if registered, may be the STS substitute the ornamentals industry has been seeking (Nell, 1992). At minute concentrations (20 $nl·liter^{-1}$), 1-MCP provided as much protection as STS, preventing ethylene-induced bud and flower abscission, leaf abscission, and flower senescence. Because the concentration required is so small, commercial application could be in the greenhouse before packing, in the transportation vehicle, or in the storage area. STS is usually effective at a concentration close to the one at which it causes phytotoxicity. We have observed no phytotoxic symptoms of 1-MCP, even at 500 $nl·liter^{-1}$. This new chemical has been patented and is commercially developed under the name EthylBloc by an American company.

To perform basic research on the mode of action of 1-MCP, we investigated its effects on symptoms of cellular senescence in petunia (*Petunia hybrida*) flowers following exposure to ethylene. Cut petunia flowers exposed to ethylene wilted sooner than their untreated counterparts. This effect was abolished by a 6 h pretreatment with 1-MCP. The ethylene treatment caused rapid decreases in petal fresh weight and total protein content, accompanied by higher electrolyte leakage, lower fluidity of membrane lipids, and reduced membrane protein content (Serek et al., 1995). When the flowers were treated with 1-MCP prior to the ethylene treatment, ethylene had none of these effects, indicating that the action of 1-MCP is at a very early stage in the regulatory cascade leading from the perception of ethylene to expression of its effects in flower senescence.

Alcohols

A number of workers have reported that alcohols, particularly ethanol, have antiethylene properties. Wu and colleagues (1992), for example, showed that keeping carnations in a vase solution containing 4 percent ethanol considerably extended flower longevity. Their analysis suggested that the ethanol was acting primarily by reducing the binding of ethylene to its binding site.

Abscisic Acid

Abscisic acid plays an important role in plant dormancy and is a key component of the regulation of plant-water relations, considered to be primarily responsible for closing the stomata (Radin and Ackerson, 1982). When potted chrysanthemum plants wilted, ABA content of the leaves rose rapidly (Cornish et al., 1985), and ABA treatment of potted chrysanthemum plants dramatically reduced stomatal aperture and water loss (see Figure 5.1).

Although the ABA-treated plants lasted three to five days longer than the untreated control, these authors considered the treatment to be of scientific rather than commercial interest. Kohl and Rundle (1972) treated cut roses with ABA and found a similar reduction in transpiration. In the case of roses, the treatment had the undesirable side effect of accelerating flower senescence (perhaps through stimulated ethylene production) (Zacarias and Reid, 1990). Thus, although abscisic acid appears to be capable of reducing postharvest water loss in potted plants, more research is required to determine whether these findings have any practical implications.

Gibberellins and Inhibitors of Gibberellin Biosynthesis

Gibberellins, primarily regulators of cell elongation, are important in ornamental horticulture largely from the perspective of the need to inhibit their action to obtain the desired height-diameter ratio for pot-

FIGURE 5.1. Effect of ABA Treatment on the Rate of Water Loss from Potted Chrysanthemum Plants

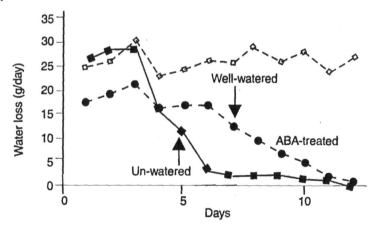

Source: Redrawn from the data of Cornish et al., 1985.

Note: Arrows mark the onset of leaf wilting.

ted flowering plants. Inhibitors of gibberellin synthesis and action, such as Alar, cycocel (CCC), and PP333 (Bonzai) are routinely used in commerce for this purpose. In some cases, the use of these regulators has undesirable effects on the postharvest life of the plants, accelerating postharvest yellowing and reducing flower life (Han, 1997).

An important commercial use of gibberellins is in preventing postharvest leaf yellowing in monocotyledonous cut flowers, such as lilies and alstroemeria. Commercial preservatives containing gibberellin have been formulated for these crops. In their study of cut statice stem pieces, Steinitz and colleagues (1980) found that immersion in high concentrations of GA_3 was even more effective than 6-BA in reducing loss of chlorophyll after harvest. An intriguing observation that has not resulted in any practical application is the delaying effect of gibberellins on senescence in isolated rose petals (Goszczynska et al., 1990). It may be that in roses, as in carnations (Saks, Staden, and Smith, 1992), gibberellins act to modify the expression of the normal senescence pathway.

Other Growth Regulators

Researchers have studied the properties of a range of other chemicals with plant hormone or growth regulator activity. The antisenescence effects of the polyamines, and the various effects of jasmonic acid and methyl jasmonate, are the most notable of these. Although intriguing findings have been reported (Halaba and Rudnicki, 1989), none of these effects has been unequivocally proven to be a natural regulatory process, nor have these compounds proved to be of practical commercial value in the postharvest life of ornamentals. Other workers have reported beneficial effects of various free-radical scavengers (Baker, Lieberman, and Anderson, 1978), and recently, a report was made of delayed "senescence" in snapdragons with lysophosphatidylethanolamine (LPE) (Kaur and Palta, 1997). This latter paper serves as a caution to those who might misinterpret the effects of growth regulators as being on senescence. The article provided a cover photograph for the journal in which it appeared. To an experienced observer, it is clear that the effects these researchers were observing were on water relations of the cut snapdragon stem, not on flower senescence, as stated in the article. It seems probable, in this case, that the application of a lipid material reduced water loss from the inflorescence, thereby preventing the negative effect on its water relations of bacterial growth in the vase solution (no biocide was provided).

Ethylene-Independent Senescence

The discovery of new methods to protect flowers and plants against ethylene action is very useful for ethylene-sensitive flowers. However, for a large group of plants, these methods will not be effective (Lukaszewski and Reid, 1989). These are the ethylene-insensitive flowers such as the geophytes (*Tulipa, Iris, Freesia,* and *Gladiolus*) and the composites (*Dendranthema, Gerbera*). Little has been done to examine the physiological basis of senescence in ethylene-insensitive flowers, which is surprising, since this type has a wide range of beautiful flowers, which could be valuable commercial products, were their postharvest life improved.

We performed a physiological study with ethylene-insensitive gladiolus flowers that offers a very interesting model system for studies of flower senescence (Serek, Jones, and Reid, 1994a). Individual florets provide a graded series of stages of development and senescence in an identical genetic and environmental background. In our study, we examined the role of ethylene in floret senescence in gladiolus flowers by determining ethylene production and respiration of individual florets during development and senescence. We also explored the effects of exogenous ethylene, STS, and sucrose treatments on opening and senescence of florets in a range of commercial cultivars. The studies confirmed the hypothesis that ethylene is not involved in senescence of gladiolus florets. STS and sucrose had similar, but not synergistic, effects on spike development, and we hypothesized that STS overcomes the effects of stress ethylene production resulting from low carbohydrate or cold stress in developing florets.

In studying the senescence of gladiolus florets, we fed them with cycloheximide (CHI), an inhibitor of protein synthesis (Jones et al., 1994). Preventing protein synthesis doubled the life of individual florets, indicating that the senescence process is regulated at the molecular level. Similar effects were obtained when iris and tulip flowers were fed CHI (see Photo 5.11).

Recent findings on the molecular basis of senescence of daylily flowers, which are also ethylene-insensitive, have provided tools for conducting basic investigations of this senescence process (Valpuesta et al., 1995). We hope that such investigations will lead to longer-lived iris in the same way that basic studies of ethylene action in plants has led to the development of 1-MCP and the potential for easily extending the life of ethylene-sensitive cut flowers and potted plants.

PHOTO 5.1. The effect of a one-hour exposure of 'Martha Washington' *Pelargonium* flowers to 1 ppm ethylene.

PHOTO 5.2. The optimum storage temperature for most cut flowers is 0°C. Flowers kept at 10°C deteriorate three times faster than those kept at 0°C.

PHOTO 5.3. *Poinsettia* plants are sensitive to low-temperature storage. After three days at 0°C, the plants on the right show typical symtpoms of chilling injury.

PHOTO 5.4. Effect of horizontal orientation during marketing on the quality of *Kalanchoe* plants.

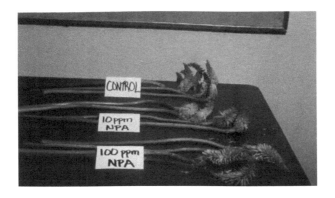

PHOTO 5.5. The gravitropism of spikes of *Kniphofia* (top) and *Antirrhinum* (bottom) can be overcome by pulsing the freshly harvested flowers with 100 ppm naphthyl pthalamic acid.

PHOTO 5.6. Effect of ethylene on epinasty in Poinsettia. Plants were treated with 0, 0.5, 2, and 15 ppm ethylene for two days. Top—intact plants; Bottom—plants with bract and leaf blades removed.

PHOTO 5.7. STS application prevents ethylene-induced abscission of flowers, leaves, and buds in miniature rose plants ('Victory Parade'), but not leaf yellowing. A combined treatment with 6-benzyladenine and STS provides plants with much superior postharvest characteristics.

PHOTO 5.8. Application of 4MM STS to *Schlumbergera* plants overcomes ethylene-induced abscission of florets and buds.

PHOTO 5.9. Diazocyclopentadiene, designed as a photo-affinity label for the ethylene binding site, proves to be a very effective inhibitor of ethylene action. Miniature rose plants treated with this material were as resistant to ethylene- induced abscission as plants treated with STS.

PHOTO 5.10. Rapid wilting of *Phalaenopsis* orchid flowers can result from ethylene exposure, or from emasculation (removal of the pollinia). Pretreatment with 20 ppb 1-MCP for six hours prevented the wilting caused by emasculation.

PHOTO 5.11. Effect of a pulse treatment with cycloheximide on the life of tulip flowers.

REFERENCES

Baker, J.E., M. Lieberman, and J.D. Anderson (1978). Inhibition of ethylene production in fruit slices by a rhizobitoxine analog and free radical scavengers. *Plant Physiology* 61:886-888.

Beyer, E.M. Jr. (1976a). Silver ion: A potent antiethylene agent in cucumber and tomato. *HortScience* 11:195-196.

Beyer, E.M. Jr. (1976b). A potent inhibitor of ethylene action in plants. *Plant Physiology* 58:268-271.

Beyer, E.M. Jr. (1978). Method for overcoming the antiethylene effects of Ag^{+1}. *Plant Physiology* 62:616-617.

Borochov, A. and W.R. Woodson (1989). Physiology and biochemistry of flower petal senescence. *Horticulture Reviews* 11:15-43.

Cameron, A.C. and M.S. Reid (1981). The use of silver thiosulfate anionic complex as a foliar spray to prevent flower abscission of zygocactus *Schlumbergera truncata*. *HortScience* 16:761-762.

Cornish, K., A.I. King, M.S. Reid, and J.L. Paul (1985). Role of ABA in stress-induced reduction of water loss from potted chrysanthemum plants. *Acta Horticulturae* 167:381-386.

Fujino, D.W., M.S. Reid, and S.F. Yang (1981). Effects of amino-oxyacetic acid on postharvest characteristics of carnation. *Acta Horticulturae* 113:59-64.

Gan, S. and R.M. Amasino (1995). Inhibition of leaf senescence by autoregulated production of cytokinin. *Science* 270:1986-1987.

Gan, S. and R.M. Amasino (1997). Making sense of leaf senescence. *Plant Physiology* 113:131-319.

Gane, R. (1934). Production of ethylene by some ripening fruits. *Nature* 134:1008.

Goeschl, J.D., Rappaport, L., and H.K. Pratt (1966). Ethylene as a factor regulating the growth of pea epicotyls subjected to physical stress. *Plant Physiology* 41:877-884.

Goszczynska, D.M., N. Zieslin, Y. Mor, and A.H. Halevy (1990). Improvement of postharvest keeping quality of 'Mercedes' roses by gibberellin. *Plant Growth Regulation* 9:293-303.

Halaba, J. and R.M. Rudnicki (1989). Jasmonate stimulates ethylene emanation by carnation flowers. *Acta Horticulturae* 251:65-66.

Han, S.S. (1997). Preventing postproduction leaf yellowing in Easter lily. *Journal of American Society for Horticultural Science* 122:869-872.

Harkema, H., M.W.C. Dekker, and M.L. Essers (1991). Distribution of amino-oxyacetic acid in cut carnation flowers after pretreatment. *Scientia Horticulturae* 47:347-353.

Jones, R.B., M. Serek, C.L. Kuo, and M.S. Reid (1994). The effect of protein synthesis inhibition on petal senescence in cut bulb flowers. *Journal of American Society for Horticultural Science* 119:1243-1247.

Joyce, D.C., M.S. Reid, and R.Y. Evans (1990). Silver thiosulfate prevents ethylene-induced abscission in holly and mistletoe. *HortScience* 25:90-92.

Kaur, N. and J.P. Palta (1997). Postharvest dip in a natural lipid, lysophosphatidylethanolamine, may prolong vase life of snapdragon flowers. *HortScience* 32:888-890.

Kohl, H.C. and D.L. Rundle (1972). Decreasing water loss of cut roses with abscisic acid. *HortScience* 7:249.

Lukaszewski, T.A. and M.S. Reid (1989). Bulb-type flower senescence. *Acta Horticulturae* 261:59-62.

Mor, Y., M.S. Reid, and A.M. Kofranek (1984). Pulse treatments with silver thiosulfate and sucrose improve the vase life of sweet peas. *Journal of American Society for Horticultural Science* 109:866.

Neljubow, D.N. (1901). Über die horizontale Nutation der Stengel von *Pisum sativum* und einiger anderen Pflanzen. *Beihefte zum Botanischen Centralblatt* 10:128-139.

Nell, T.A. (1992). Taking silver safety out of the longevity picture. *GrowerTalks* June: 41-42.

Nichols, R. (1966). Ethylene production during senescence in carnation (*Dianthus caryophyllus*). *Journal of American Society for Horticultural Science* 41:279-290.

Nichols, R. (1977). Sites of ethylene production in the pollinated and unpollinated senescing carnation (*Dianthus caryophyllus*) inflorescence. *Planta* 135:155-159.

Radin, J.W. and R.C. Ackerson (1982). Does abscisic acid control stomatal closure during water stress? *What's New in Plant Physiology* 12:9-12.

Reid, M.S. (1985). Ethylene and abscission. *HortScience* 20:45-50.

Reid, M.S. and R.Y. Evans (1988). Preparation and use of STS for cut rose flowers. *Flower and Nursery Report for Commercial Growers, University of California, Berkeley,* Spring 1988, p. 3.

Reid, M.S., R.Y. Evans, and L.L. Dodge (1989). Ethylene and silver thiosulfate influence opening of cut rose flowers. *Journal of American Society for Horticultural Science* 114:436-440.

Reid, M.S., Y. Mor, and A.M. Kofranek (1981). Epinasty of poinsettias—The role of auxin and ethylene. *Plant Physiology* 67:950-952.

Reid, M.S., J.L. Paul, M.B. Farhoomand, A.M. Kofranek, and G.L. Staby (1980). Pulse treatments with the silver thiosulfate complex extend the vase life of cut carnations. *Journal of American Society for Horticultural Science* 105:25-27.

Reid, M.S., and M.J. Wu (1992). Ethylene and flower senescence. *Plant Growth Regulation* 11:37-43.

Saks, Y., J. Van Staden, and M.T. Smith (1992). Effect of gibberellic acid on carnation flower senescence: Evidence that the delay of carnation flower senescence by gibberellic acid depends on the stage of flower development. *Plant Growth Regulation* 11:45-51.

Saltveit, M.E., D.M. Pharr, and R.A. Larson (1979). Mechanical stress induces ethylene production and epinasty in poinsettia *Euphorbia pulcherrima* cultivars. *Journal of American Society for Horticultural Science* 104:452-455.

Serek, M. and A.S. Andersen (1993). AOA and BA influence on floral development and longevity of potted 'Victory Parade' miniature rose. *HortScience* 28: 1039-1040.

Serek, M., R.B. Jones, and M.S. Reid (1994). Role of ethylene in opening and senescence of *Gladiolus* sp. flowers. *Journal of American Society for Horticultural Science* 119:1014-1019.

Serek, M. and M.S. Reid (1993). Anti-ethylene treatments for potted Christmas cactus—Efficacy of inhibitors of ethylene action and biosynthesis. *HortScience* 28:1180-1181.

Serek, M., E.C. Sisler, and M.S. Reid (1993). Commercial prospects for moderating the effects of ethylene in potted flowering plants. *Proceedings Australasian Postharvest Conference, Gatton, Australia.* Lawes, Queensland, Australia: The University of Queensland, Gatton College, pp. 423-426.

Serek, M., E.C. Sisler, and M.S. Reid (1994a). A volatile ethylene inhibitor improves the post-harvest life of potted roses. *Journal of American Society for Horticultural Science* 119:572-577.

Serek, M., E.C. Sisler, and M.S. Reid (1994b). Novel gaseous ethylene binding inhibitor prevents ethylene effects in potted plants. *Journal of American Society for Horticultural Science* 119:1230-1233.

Serek, M., E.C. Sisler, and M.S. Reid (1995). Effects of 1-MCP on the vase life and ethylene response of cut flowers. *Plant Growth Regulation* 16:93-97.

Serek, M., E.C. Sisler, and M.S. Reid (1996). Ethylene and the postharvest performance of miniature roses. *Acta Horticulturae* 424:145-149.

Serek, M., G. Tamari, E.C. Sisler, and A. Borochov (1995). Inhibition of ethylene-induced cellular senescence symptoms by 1-methylcyclopropene, a new inhibitor of ethylene action. *Physiologia Plantarum* 94:229-232.

Shih, C.Y., E.B. Dumbroff, and J.E. Thompson (1989). Identification of a naturally occurring inhibitor of the conversion of 1-aminocyclopropane-1-carboxylic acid to ethylene by carnation microsomes. *Plant Physiology* 89:1053-1059.

Sisler, E.C. (1977). Ethylene activity of some pi acceptor compounds. *Tobacco Science* 21:43-45.

Sisler, E.C. (1991). Ethylene-binding components in plants. In *The Plant Hormone Ethylene,* A.K. Mattoo and J.C. Suttle (Eds.). Boca Raton, FL: CRC Press, pp. 81-99.

Sisler, E.C. and S.M. Blankenship (1993). Effect of diazocyclopentadiene on tomato ripening. *Plant Growth Regulation* 12:155-160.

Sisler, E.C., E. Dupille, and M. Serek (1996). Effect of 1-methylcyclopropene and methylenecyclopropane on ethylene binding and ethylene action on cut carnations. *Plant Growth Regulation* 18:79-86.

Sisler, E.C. and N. Lallu (1994). Effect of diazocyclopentadiene (DACP) on tomato fruits harvested at different ripening stages. *Postharvest Biology and Technology* 4:245-254.

Sisler, E.C., M.S. Reid, and D.W. Fujino (1983). Investigation of the mode of action of ethylene in carnation senescence. *Acta Horticulturae* 141:229-234.

Sisler, E.C. and M. Serek (1997). Inhibitors of ethylene responses in plant at the receptor level: Recent developments. *Physiologia Plantarum* 100:577-582.

Sisler, E.C., M. Serek, and E. Dupille (1996). Comparison of cyclopropene, 1-methylcyclopropene, and 3,3-dimethylcyclopropene as ethylene antagonists in plants. *Plant Growth Regulation* 18:169-174.

Sisler, E.C. and S.F. Yang (1984). Anti-ethylene effects of *cis*-2-butene and cyclic olefins. *Phytochemistry* 23:2765-2768.

Staby, G.L., R.M. Basel, M.S. Reid, and L.L. Dodge (1993). Efficacies of commercial anti-ethylene products for fresh cut flowers. *HortTechnology* 3:199-202.

Staby, G.L., B.A. Eisenber, J.W. Kelly, M.P. Bridgen, and M.S. Cunningham (1981). The role of ethylene in poinsettia petiole epinasty. *Florists' Reviews,* June 18, 168:44-45.

Steinitz, B., A. Cohen, and B. Leshem (1980). Factors controlling the retardation of chlorophyll degradation during senescence of detached statice (*Limonium sinuatum*) flower stalks. *Zeitschrift für Pflanzenphysiologie* 100:343-349.

Tjosvold, S.A., M.J. Wu, and M.S. Reid (1994). Reduction of postproduction quality loss in potted miniature roses. *HortScience* 29:293-294.

Valpuesta, V., N.E. Lange, C. Guerrero, and M.S. Reid (1995). Up-regulation of a cysteine protease accompanies the ethylene-insensitive senescence of daylily (*Hemerocallis*) flowers. *Plant Molecular Biology* 28:575-582.

Van Doorn, W.G. and M.S. Reid (1992). Role of ethylene in flower senescence of *Gypsophila paniculata* L. *Postharvest Biology and Technology* 1:265-272.

Veen, H. (1983). Silver thiosulphate: An experimental tool in plant science. *Scientia Horticulturae* 20:211-224.

Veen, H. and S.C. Van de Geijn (1978). Mobility and ionic form of silver as related to longevity in cut carnations. *Planta* 140:93-96.

Wang, C.Y., J.E. Bakker, R.E. Hardenburg, and M. Lieberman (1977). Effects of two analogs of rhizobitoxine and sodium benzoate on senescence of [cut] snapdragons. *Journal of American Society for Horticultural Science* 102:517-520.

Woltering, E.J., H. Harkema, M.A. Maclaine Pont, and P.C.H. Hollman (1987). Amino-oxyacetic acid: Analysis and toxicology. *Acta Horticulturae* 216: 273-279.

Woltering, E.J., D. Somhorst, J.G. Beekhuizen, and W.T.J. Spekking (1991). Ethylene biosynthesis, carbohydrate metabolism and phenylalanine ammonia-lyase activity in gravireacting *Kniphofia* flower stalks. *Acta Horticulturae* 298: 99-109.

Woltering, E.J. and W.G. Van Doorn (1988). Role of ethylene in senescence of petals—Morphological and taxonomical relationships. *Journal of Experimental Botany* 39:1605-1616.

Woodson, W.R., A.S. Brandt, H. Itzhaki, J.M. Maxson, K.Y. Park, and H. Wang (1993). Regulation and function of flower senescence-related genes. *Acta Horticulturae* 336:41-46.

Wu, M.J., L. Zacarias, M.E. Saltveit, and M.S. Reid (1992). Alcohols and carnation senescence. *HortScience* 27:136-138.

Yang, S.F., and N.E. Hoffman (1984). Ethylene biosynthesis and its regulation in higher plants. *Annual Review of Plant Physiology* 35:155-189.

Zacarias, L. and M.S. Reid (1990). Role of growth regulators in the senescence of *Arabidopsis thaliana* leaves. *Physiologia Plantarum* 80:549-554.

Chapter 6

Manipulating Fruit Development and Storage Quality Using Growth Regulators

Susan Lurie

INTRODUCTION

The use of plant growth regulators (PGRs) during various stages of fruit development is a well-established practice. By the end of the 1940s, a number of production technologies had been developed utilizing PGRs. F. G. Gustafson (1936) was the first to demonstrate that a specific chemical could be used to achieve fruit set and full-organ development without pollination. In addition, hundreds of studies dealing with the improvement of fruit set of all kinds of vegetables, berries, and tree fruits had been reported (Looney, 1997). Another early application of hormones to tree fruit horticulture was the use of naphthaleneacetic acid (NAA), a synthetic auxin, or its amide salt to reduce fruit set of apple. NAA was tried by Schneider and Enzie (1944), and within a few years, NAA applied after the bloom period to reduce fruit set (even of pollinated flowers) was an established practice. The same auxin was found early on to be efficacious in reducing preharvest drop of apples. Gardner and colleagues (1940) developed this technology for apples, and an early review of this field cited sixty-nine references to the chemical control of abscission, mainly of apples (Avery and Johnson, 1947).

These first applications of PGRs to horticulture center around auxin, and chemicals with auxinlike activity, because it was the first plant hormone isolated. However, cytokinins, gibberellins, and ethylene have all proved useful, both in the past and in the present, for improving the development of fruits. In addition, in the past decade, other PGRs have entered into horticultural use. These include antigibberellins such as paclobutrazol and uniconazole, as well as PGRs based on natural products such as jasmonic acid.

Just as the chemicals in use to aid fruit development are many and varied, so are the uses to which they are put. They have been used to control fruit set, produce seedless fruits, aid in fruit thinning, increase fruit size, hasten or delay ripening and maturity, and improve post-harvest keeping quality. This chapter cannot comprehensively cover all of these areas but will instead try to give an overview of the different stages of fruit development and the various PGRs that are used at each stage.

THINNING

If fruits are clustered too close together, they will not develop to an adequate size. Most tree fruits set too many fruits and need to be thinned to a proper fruit-leaf area ratio to enable larger fruit growth. Hand thinning is time-consuming, labor intensive, and often the single greatest cost of production. Therefore, much research has been invested in efforts to develop reliable chemical thinning regimes. In addition, many tree crops show biennial bearing, whereby one year the tree will give a very heavy crop and the following year a light crop. Thinning can often reduce or eliminate this problem, along with enhancing fruit size, shape, color, and overall quality.

Many PGRs, as well as other chemicals, have been used to thin fruits. The first chemical found to be effective, which is still in use, was 4,6-dinitro-ortho-crysylate (DNOC), a caustic and nonselective weed killer applied during flowering. Since the action of this chemical prevents pollination by burning the stigmas, it cannot be used for postbloom thinning (Weaver, 1972). Postbloom thinning is preferable because it provides the grower with an opportunity to evaluate the degree of fruit set before applying sprays. Carbamates, which are also used as fungicides or insecticides, can affect fruit set and thinning (Veinbrants, 1975). As these sprays are applied about the same time as chemical fruit thinning sprays, they may affect fruit set. Two carbamates, carbaryl (also known as Sevin) and oxamyl (also known as Vydate), have been found particularly successful for fruit thinning (see Table 6.1).

In the area of thinning, many PGRs have been found useful, including auxin, cytokinin, and gibberellin, in combination or individually, and ethylene-releasing compounds such as ethephon. Table 6.1 shows compounds used chemically for apple thinning. Apple thinning has been invested in heavily because it is an important crop worldwide and most cultivars require it.

Some cultivars, such as spur-type 'Delicious', are more difficult to thin than others, and an aggressive thinning program is usually required (Williams and Edgerton, 1981). Carbaryl and NAA are the primary chemical thinners used on 'Delicious' in the northeastern United States (Forshey, 1987). Carbaryl is a relatively mild thinner and is frequently not adequate when used alone (Williams and Edgerton,

Table 6.1. Chemicals That Have Been Used to Thin Apples

Common name	Chemical name	Trade name
Ethephon	(2-chloroethyl) phosphonic acid CEPA	Ethephon, Ethrel
Silaid, Alsol	(2-chloroethyl) methylbis (phenylmethoxy) silane; (2-chloroethyl) tris (2-methoxy) silane	Silaid, Alsol, Etacelasil
DNOC	Sodium 4,6-dinitro-o--cresylate; 4,6-dinitro-o-cresol	Elgetol, Dinitro-dry
NAA	Naphthaleneacetic acid	Fruitone-N, Fruit Fix-800, Fruit Fix-200, Fruit Set, Stafast, Kling-Tite
NAAm	Naphthalene acetamide	Amide-Thin W. Anna-Amide
Carbaryl	1-naphthyl N-methyl carbamate	Sevin
Oxamyl	Methyl N'N-dimethyl-N-{(methyl carbamoyl)oxy}-1-thiooxamimidate	Vydate
Cytokinin + gibberellin	N-(phenylmethyl)-H-purine 6-amine and $GA_4 + GA_7$	Accel, Promalin
Silvex, Fenoprop	2-(2,4,5-trichlorophenoxy) propanoic acid, 2,4,5-TP	Fruitone T

1981). Its usefulness is further limited because it is toxic to some mite predators important in integrated pest management (Hislop and Prokopy, 1981). NAA is a more potent thinner; however, when applied at concentrations that thin adequately, it may cause pygmy fruit formation or may overthin (Rogers and Williams, 1977; Bound, Jones, Koen, Oakford, et al., 1991).

The ethylene-releasing compound ethephon has shown thinning properties on several apple cultivars (Knight, 1980; Jones, Koen, and Meredith, 1983; Jones et al., 1989; Knight et al., 1987; Koen, Jones, and Longley, 1988). The cytokinins benzyladenine (BA) and N-(2-chloro-4-pyridyl)-N'-phenylurea (CPPU) also thin apple trees (Elfving, 1989; Greene, 1989; Greene and Autio, 1989; Greene, Autio, and Miller, 1990; Bound, Jones, Koen, and Oakford, 1991; Byers and Carbaugh, 1991; Elfving and Cline, 1993a). Thidiazuron has been reported to have cytokinin-like activity in apple and other species (Wang, Steffens, and Faust, 1986; Fellman, Read, and Hosier, 1987) and may be effective as a thinner at low concentrations (Elfving and Cline, 1993b). Cytokinins and ethephon can also affect vegetative growth, nutrient concentrations, and crop load (Elfving, 1984; Miller, 1988). Cytokinins have also been combined with carbaryl and NAA to thin 'McIntosh' apples (Greene and Autio, 1989).

In stone fruits, gibberellin has been investigated for thinning. Gibberellic acid (GA_3) inhibits flower bud differentiation in deciduous fruit (Hull and Lewis, 1959; Bradley and Crane, 1960; Liu et al., 1989). Gibberellin (GA) sprays applied to stone fruits (peach and apricot) reduced the following season's flowering and increased fruit size compared to hand-thinned and nonthinned trees, as well as eliminating the need for hand thinning in some seasons (Southwick and Yeager, 1995; Southwick et al., 1995; Southwick, Yeager, and Zhou, 1995).

Grapes are another crop that requires thinning. The incidence of rot in grape species or varieties that produce tight clusters is very high because tight clusters dry very slowly after rain. In addition, tight clusters also cause berries within the cluster to be crushed, thereby providing an excellent media for decay-causing organisms. Gibberellin dips of the clusters are used to elongate the bunches, thereby removing the need for thinning (Weaver and Pool, 1965). Timing and concentration are important to have the optimum effect on the berries.

FRUIT DEVELOPMENT

Fruit growth typically follows two distinct growth curves. One type is a smooth sigmoid curve that is exhibited by many tree crops, including apples and pears. The second type of growth curve is represented by a double sigmoid that can be viewed as two successive sigmoid curves. Blueberries, grapes, figs, and the stone fruits show this type of growth pattern. This growth curve has two periods of rapid growth separated by an intermediate period during which little or no growth occurs, but, instead, lignification of the endocarp takes place. The use of PGRs to influence fruit growth has become important in agriculture today because they have the ability to increase fruit size and improve color and shape, thereby increasing marketability. A number of PGRs can influence the final size of the fruit, each by a different mechanism.

Gibberellins in various formulations may be the most common of the PGRs used for this purpose. Gibberellin enhances cell enlargement, thereby increasing size. It also has the effect of delaying ripening, so that the fruit has a longer time to develop and therefore reaches a larger size. When the treatment or treatments are given during the fruit growth will determine which of these two effects will be primary. Cherries are generally sprayed with gibberellin three to four weeks before harvest (Looney and Lidster, 1980; Facteau, Rowe, and Chestnut, 1985). The sprayed fruits are both larger and firmer than the unsprayed cherries (Looney and Lidster, 1980; Facteau, 1982). Similar responses have been found in other stone fruits, such as apricots (Southwick and Yeager, 1995) and nectarines (Zilkah et al., 1997). Persimmons are normally given a gibberellin spray to delay their ripening (Gross et al., 1984), but earlier sprays were found to enhance the final fruit size (Hasegawa et al., 1991). In citrus, gibberellin sprays are used to delay rind pigmentation and prevent the development of rind blemishes during the harvest season. In most cases, the late application does not affect fruit size, though an increase has been reported in a few cases (El-Zeftawi, 1980a; Lima and Davies, 1984).

The fruits that receive the greatest number of gibberellin treatments are the various cultivars of seedless grape. The time of gibberellin applications profoundly affects the ability to promote enlargement of fruits and berry shape (Christodoulou, Weaver, and Pool, 1968; Zuluaga, Lumelli, and Christensen, 1968). Gibberellin treatments made at the time of full bloom elongate the berries, and later applications produce larger berries (Christodoulou, Weaver, and Pool, 1968). The later applications of gibberellic acid may also enhance the soluble solids of the grapes at harvest (Harrell and Williams, 1987). The enhancement of fruit growth promoted by gibberellins in seeded cultivars of grapes is generally minimal. An increase in size of seeded varieties of grape is generally associated with a lower number of seeds found within the fruit, suggesting that endogenous levels of gibberellin are affecting fruit size, and when these levels are high enough, exogenous applications have no effect.

Recently, much attention has been paid to PGRs that inhibit gibberellin synthesis but, paradoxically, often increase fruit size. The original use of these compounds, mainly paclobutrazol, also known as PP333 or Cultar, and uniconazole, also known as Magic, was to control tree vigor. Deciduous orchards are increasingly being planted as high-density orchards where economic success depends on manipulation of the trees to crop early and to control growth once trees have reached the desired size. Excessive tree vigor can reduce flower bud formation and fruit set, resulting in reduced fruit quality. The use of PGRs to control tree growth can lead to a small, balanced canopy (Curry and Jones, 1990).

Because these two PGRs, paclobutrazol and uniconazole, inhibit endogenous gibberellin synthesis, they reduce internode length in the branches and can also decrease the number of nodes per shoot (Blanco, 1988). This affects assimilate partitioning, with more assimilates being available to fruit, thereby increasing fruit size (Martin, Yoshikawa, and LaRue, 1987). However, as with many PGRs, timing and concentration are critical (Davis, Steffens, and Sankhla, 1988). Undesirable side effects of concentrations of paclobutrazol used to control apple tree growth have been reduced fruit size (Elfving and Proctor, 1986; Greene, 1986; Miliou and Sfakiotakis, 1986; Prive, Elfving, and Proctor, 1989), a change in fruit shape (Elfving et al., 1987; Jones et al., 1988; El-Khoreiby, Unrath, and Lehman, 1990), and reduced fruit soluble-solids content, probably due to decreased leaf area (Curry and Williams, 1986; Greene, 1986; Sansavini et al., 1986; Elfving et al., 1990). Paclobutrazol can also increase fruit thinning and fruit russeting (Church, Copas, and Williams, 1984; Volz and Knight, 1986; El-Khoreiby, Unrath, and Lehman, 1990; Steffens, Jacobs, and Engelhaupt, 1993) if applied early in the postbloom period. However, most of these undesirable effects can be avoided if the application is delayed until later in the season, or if several sprays of reduced concentration are given instead of one spray (Miliou and Sfakiotakis, 1986; Quinlan and Richardson, 1986; Curry and Jones, 1990; Greene, 1991).

In the stone fruits, nectarines, peaches and plums, paclobutrazol, either by a soil treatment in the previous fall or in the spring, has been found to increase fruit size at harvest (Martin, Yoshikawa, and LaRue, 1987; Blanco, 1990; George et al., 1995; Lurie et al., 1997). In cherries, the results are more equivocal. In some studies, total tree yields were reduced while fruit size was increased (Webster, Quinlan, and Richardson, 1986; Looney and McKellar, 1987). Paclobutrazol is not readily metabolized and can influence tree physiology for more than one season. In some cases, the inhibition of stem growth can continue for two or three years. Often with cherries fruit size increases the first year. However, due to restricted stem growth, the fruit set in later seasons will be borne on mature spurs where fruit quality is reduced (Looney and McKellar, 1987; Facteau and Chestnut, 1991).

Orchards of subtropical fruit, such as avocados and mangoes, also utilize paclobutrazol or uniconazole to control tree vigor and increase fruit size and yield (Whiley et al., 1991; Kurian and Iyer, 1993a, b). In these orchards, the treatment is often given as a foliar spray, which sometimes allows a lower concentration to be used.

Cytokinins can be used to promote fruit growth, but only at early times of fruit development. Cytokinins enhance cell division; thus, if applied at the early part of the sigmoidal growth curve (or in the first part of the double sigmoid) when cell division is occurring, they may lead to more cells per fruit. This can result in larger fruit at maturity. CPPU influences cell division and morphogenesis in various crops. It has been shown to increase final fruit size in apples (Tartarini, Sansavini, and Ventura, 1993), kiwifruit (Iwahori, Tominaga, and Yamasaki, 1988; Patterson, Mason, and Gould, 1993), and grapes (Ben Arie et al., 1997).

FRUIT DROP

The predominant PGRs in the regulation of premature fruit abscission, or preharvest drop, are auxins and ethylene. Many of the early observations on fruit storage were in conjunction with the synthetic auxins that were used for the control of premature fruit drop, such as NAA and 2,4,5-trichlorophenoxypropionic acid (2,4,5-TP). In many instances, these compounds were found to promote red color development and fruit ripening in apples. For example, Southwick, Demoranville, and Anderson (1953) showed that 2,4,5-TP applied three weeks before harvest of 'McIntosh' apples increased the proportion of fruit in the extra-fancy grade but decreased fruit firmness. NAA is generally slightly less effective than 2,4,5-TP in promoting red color development and softening, and both were found to have no adverse effects on the storage quality of 'Cox' apples (Sharples, 1973). Today, NAA, under trade names such as Fruit Fix 200, Fruit Set, Stafast, and others, is commercially used for the control of apple and pear preharvest drop. 2,4,5-TP is also known as silvex or fenoprop and by such trade names as Fruitone T and has been applied as a foliar spray two weeks prior to harvest for most varieties of apples. However, because of concern about a manufacturing contaminant, 2,3,7,8-tetrachlorobenzo-p-dioxin, this compound is no longer being used in the United States, and its use is decreasing elsewhere.

Daminozide, under the trade name Alar, has been shown to be effective in reducing harvest drop in apples when applied as a preharvest spray. It has also been reported to have a number of other beneficial effects, such as reducing water core and storage scald, maintaining fruit firmness, and increasing fruit color (Arteca, 1996). However, at present, it has been removed from commercial use due to potential health risks. Instead, recent studies with the PGR aminoethoxyvinyl glycine (AVG), which inhibits ethylene biosynthesis, have been reexamined (Bangerth, 1978). This compound, now marketed under the trade name Retain, can give many of the beneficial effects that Alar formerly did. Retain delayed maturity, allowing for later harvest; maintained fruit firmness; and reduced superficial scald in storage (Watkins et al., 1997). However, unlike Alar, it reduced red color development on apples, which is undesirable. This is a recently introduced compound that will require more testing.

FRUIT RIPENING

The manipulation of fruit ripening is of major economic importance. In fact, the term climacteric refers to fruits that will ripen in response to ethylene, such as apple, banana, and avocado, and non-climacteric refers to fruits that will not ripen in response to ethylene, such as citrus, cherry, and grape (Abeles, Morgan, and Saltveit, 1992). However, even nonclimacteric fruits will respond to ethylene, though not by producing sustainable high levels of endogenous ethylene, as the climacteric fruits do. Citrus fruits are routinely de-greened by an ethylene treatment postharvest. The ethylene causes the acceleration of chlorophyll breakdown in the fruit peel, without affecting internal fruit quality.

Several lines of evidence show that ethylene is involved in the ripening of climacteric fruits. First, it has been shown that exogenous applications of ethylene will promote ripening. Second, prior to fruit ripening, endogenous levels of ethylene are very low; however, once the ripening process is initiated, there is a dramatic increase in ethylene production and subsequent ripening. Third, ripening can be retarded by hypobaric storage, treatment with inhibitors of ethylene synthesis or action, and removal of ethylene from storage rooms by scrubbing.

Present knowledge on the involvement of ethylene in the fruit-ripening process can be used in two ways. The first way is to promote faster, more uniform ripening to meet market demands and to facilitate mechanical harvesting. At present, ethephon, which releases ethylene upon breakdown, can be used to speed and promote uniform ripening in cherries, apples, boysenberries, pineapples, blueberries, coffee, cranberries, and figs (Arteca, 1996). The second way is to delay ripening to enhance the shelf life of the fruit. Increased shelf life can be accomplished by blocking ethylene biosynthesis or action or using conditions that remove ethylene from around the fruit. This latter technique will be discussed in the next section.

Ethephon can be used on fruits for different purposes that indirectly affect the rate of fruit maturation and ripening. In grapes, vigorous growth of developing canes can induce shading inside the canopy that decreases fruit quality. Shading within a canopy causes reduced soluble-solids concentration, higher malic acid, and reduced anthocyanin content (Weaver, 1975; Smart, 1982). Shading effects are usually overcome by topping the growing shoots (Lavee, 1981; Smart, 1982). However, treatment with ethephon can also be used to reduce the amount of pruning needed and increase light penetration into the canopy (Patterson and Zoecklein, 1990). More light will increase anthocyanin production, which is dependent on a certain amount of ultraviolet (UV) from sunlight and which is important for good color development in red or purple grapes. 'Barlinka' grape-vines are treated with ethephon about three weeks before harvest to improve color development. This treatment promotes more intense and even coloration of the berries and enables the crop to be har-vested in three to four pickings instead of the normal seven or eight (Blommaert, Hanekom, and Steenkamp, 1984). Giving an ethephon treatment can also aid in harvesting by accelerating the formation of the abscission layer and facilitating fruit separation and harvesting. This is of interest in wine grapes such as 'Concord', which are mechanically harvested (Morris and Cawthon, 1982). The treatments for color development and decreasing the removal force needed for harvesting are given within a month of harvest. This time of treat-ment will also hasten maturation. 'Thompson Seedless' grapevines were treated with ethephon at veràsion or with gibberellin soon after fruit set. The ethephon-treated grapes matured sixteen days earlier than did the controls, while the gibberellin treatment retarded matu-ration thirty-two days beyond the control fruits (El-Banna and Weaver, 1979). Therefore, the harvesting period of grapes can be greatly extended through the judicial use of PGRs.

Ethephon is also applied to stone fruits to hasten ripening, give more uniform fruit ripening, and allow for earlier picking. It is gener-ally used on early cultivars to enable farmers to reach the market quickly. However, in some peach and apricot cultivars, this treatment also caused the fruit to stop growing, an undesirable trait, since size is achieved mainly through cell expansion close to harvest (Ga'ash and Lavee, 1982). In addition, ethephon may cause accelerated softening of the fruit at harvest and during shelf life (Ben Arie and Guelfat-Reich, 1975). Nonetheless, with the proper dose, it is generally possi-ble to achieve the beneficial effects of ethephon without the undesir-able ones. Both plums and peaches can achieve earlier and uniform ripening without compromising size and firmness (Looney, 1972; Blommaert, Theron, and Steenkamp, 1975).

An alternative to giving ethephon, which releases ethylene as it decomposes, is to give synthetic auxins that will enhance endoge-nous ethylene production. 2,4,5-TP is often given to apricots at pit hardening both to enhance fruit size and to advance ripening (Guelfat-Reich and Ben Arie, 1975).

Ripening can also be affected by treatments that have been made to the orchard during the development of the fruit. The treatments used for thinning, fruit growth, and prevention of premature abscission can all effect ripeness of the fruit at harvest. Crop load can affect the firmness of fruit at harvest, and, therefore, adequate thinning, whether by chemicals or by hand, can be beneficial. Optimal thinning shortly after fruit set increased both size and firmness of apples at harvest (Johnson, 1994). The use of cytokinins for thinning can enhance fruit size and increase soluble solids in apples (Elfving and Lougheed, 1994). However, in kiwifruit and nectarines, CPPU will advance ripening and accelerate softening (Iwahori, Tominaga, and Yamasaki, 1988; Lurie, unpublished data).

Paclobutrazol has been found to have numerous effects on fruit ripening. In apples, spring foliar paclobutrazol sprays reduced preharvest fruit drop and increased flesh firmness, but sometimes soluble-solids content was reduced (Elfving et al., 1987, 1990). In the year following the application of low concentrations of paclobutrazol the fruits showed no residual effects, though shoot growth was still reduced. However, at high concentrations, effects on the fruit were seen the following year (Greene, 1986).

In stone fruit, soil injection or a trunk drench of paclobutrazol given to control vegetative growth was found to increase fruit yield and size of peaches and nectarines (Martin, Yoshikawa, and LaRue, 1987; Blanco, 1990). The treatments also advanced fruit maturity, and a higher percentage of the crop was harvested at the first picking. However, here, too, soluble-solids content was adversely affected by the PGR. Spring soil applications of paclobutrazol and uniconazole advanced plum fruit color development without affecting fruit firmness, soluble solids, or acidity (Lurie et al., 1997). Treatments with the PGRs increased fruit weight and size in all three seasons observed, including the third season, when no application was given.

In contrast to paclobutrazol and uniconazole, gibberellin treatments to enhance fruit growth will generally delay ripening. In citrus, which is nonclimacteric, the fruit is often stored on the tree when it is edible. This tree-stored fruit retains high internal quality, but the peel becomes senescent. Gibberellin delays changes in peel pigmentation and prevents the development of blemishes during the harvest season; therefore, it has been recommended for oranges (Coggins, 1981), lemons (El-Zeftawi, 1980a), grapefruit (El-Zeftawi, 1980b; Ferguson et al., 1984), and several mandarins, such as 'Satsuma' (Kuraoka, Iwasaki, and Ishii, 1977; Agusti, Almela, and Guardiola, 1981) and 'Clementine' (El-Otmani, M'Barek, and Coggins, 1990;

Garcia-Luis, Herrero-Villen, and Guardiola, 1992), particularly when the fruit is harvested late in the season. The maintenance of peel juvenility has an additional benefit of decreasing infestation by fruit flies, a major problem in citrus (McDonald et al., 1997).

Gibberellin is also used to delay ripening of persimmon (Ben Arie, Bazak, and Blumenfeld, 1986). The fruits are firmer at harvest and store better, whereas paclobutrazol-treated fruits ripen and are picked earlier and are designated for the fresh market. Through judicious use of these two PGRs, the harvest season for 'Triumph' persimmons, which is normally two weeks in length, can be extended to six weeks.

In stone fruits, gibberellin applied to increase fruit size will also enhance fruit firmness and delay harvest (Southwick and Yeager, 1995; Zilkah et al., 1997). In cherries, this treatment also enhances soluble-solids content (Looney and Lidster, 1980; Facteau, Rowe, and Chestnut, 1985).

STORAGE QUALITY

Citrus is the major crop for which PGRs are used for postharvest purposes. Early in the harvest season, many citrus cultivars are mature internally, but the peel has not yet developed sufficient yellow or orange color. Gassing with ethylene will accelerate chlorophyll degradation and cause carotenoid accumulation (Davies, 1986). This is generally done postharvest by holding the fruit at 20 to 30°C in rooms with 5 to 10 parts per million (ppm) ethylene for one to three days, depending on the cultivar, the growing conditions, and when in the season the fruit is picked (Grierson, Cohen, and Kitagawa, 1986). The fruit is also given a dip with a fungicide containing the auxin 2,4-dichlorophenoxy acetic acid (2,4-D). This PGR maintains the green color of the fruit button during storage and decreases stem end rots. The effect is achieved both by prevention of tissue senescence, which makes it more susceptible to pathogen invasion, and by direct inhibition of spore germination and growth (Cohen, Latter, and Barkai-Golan, 1965).

In many storage rooms, particularly with climacteric fruits, quality is maintained and softening inhibited by the removal of ethylene produced by the fruit from the storage atmosphere. This is done by passing the air during recirculation through a scrubber containing a compound that will bind the ethylene. Kiwifruit are particularly sensitive to ethylene and will soften in storage in the presence of low levels of this PGR. However, other fruits, such as apples and plums, also benefit from ethylene removal. Recently, Wills (1998) has shown that even nonclimacteric produce, such as leafy vegetables, will have its shelf life extended if no ethylene is present.

An alternative to ethylene removal is lowering the sensitivity of the fruit to its effect. A new inhibitor of ethylene action, diazocyclopentadiene, has been found to inhibit ripening in tomatoes and apples (Blankenship and Sisler, 1993). It acts by occupying the ethylene binding site so that ethylene cannot attach and elicit subsequent actions. Gibberellin may also exert some of its delay of ripening by inhibiting ethylene receptor sites, since it was found that gibberellin reduces the sensitivity of persimmon fruit to ethylene (Ben Arie et al., 1989).

Gibberellin has been found to have a number of beneficial effects on fruits in storage. In citrus, gibberellin inhibits the peel senescence without inhibiting the internal ripening of the fruit. Postharvest, the peel remains less susceptible to attack by fruit flies and also to decay problems (Greany et al., 1994). This protection against postharvest decay by gibberellin was also found in nectarine fruits (Zilkah et al., 1997). In addition, a gibberellin spray at the end of pit hardening decreased the development of physiological storage disorders in the nectarines. This effect was specific for application of gibberellin at this time of fruit development. Later application increased fruit size and firmness but did not prevent storage disorders (Zilkah et al., 1997). Gibberellin treatment can also decrease cherry pitting (Looney and Lidster, 1980). Although it enhances fruit firmness and decreases bruising, it appears that the effect on pitting is not a direct result of this increased firmness. In contrast, the PGRs that are gibberellin inhibitors, when used in high concentrations, can increase storage disorders in stone fruit. Plums treated with high levels of paclobutrazol and uniconazole can develop more internal gel breakdown in storage than untreated fruit (Lurie et al., 1997).

The effect of paclobutrazol treatments in apple orchards to control tree growth on the postharvest behavior of the fruit has been studied extensively. It was found that paclobutrazol increased calcium concentration in the fruit (Greene, 1991). Many storage disorders in apple are inversely related to fruit calcium concentration, and, therefore, fruit from paclobutrazol-treated trees had less bitter pit, cork spot, and senescent breakdown (Greene, 1991). In 'Spartan' apples, greater firmness and less internal breakdown was present after storage in paclobutrazol-treated apples (Wang and Steffens, 1987). The effects of paclobutrazol on 'McIntosh' apples have had conficting reports. In one study, paclobutrazol led to firmer fruit after storage and less core browning than in untreated apples (Elfving et al., 1987). In another study, flesh firmness was decreased and brown core increased by certain paclobutrazol treatments (Elfving et al., 1990). Therefore, the postharvest effects of this PGR cannot be generalized from one study.

A storage disorder of apples that is influenced by the fruit maturity at harvest is superficial scald. Later-harvested fruits are less susceptible to this disorder. Therefore, ethephon treatment before harvest to advance maturity can be beneficial in reducing this disorder (Greene, Lord, and Bramlage, 1977; Lurie, Meir, and Ben Arie, 1989). A combination of ethephon plus auxin (2,4,5-TP or NAA) on 'McIntosh' apples will also advance ripening without any loss of storage life (Looney, 1975; Forsyth et al., 1977). On the other hand, ethephon treatment of 'Cox's Orange Pippin' apples that advanced maturity also enhanced senescent breakdown in storage (Watkins et al., 1989). However, the benefits of ethephon treatment on a very scald-susceptible cultivar, such as 'Granny Smith', may outweigh the risks of enhancing other storage disorders.

CONCLUSIONS

The foregoing account shows that the use of PGRs is very important in modern fruit cultivation. However, since most PGRs are synthetic compounds there is always a danger of their removal from use. The scientific literature of the 1970s and 1980s was full of reports on Alar and its benefits and limitations for improving the quality of various fruits. Then a potential danger to health was found, and it was removed from registration. Replacements are still being investigated. Paclobutrazol and uniconazole are in great demand right now because of the modern orchard practice of high-density planting and their ability to control vegetative growth. However, they are long-lived compounds that, in the long run, may cause environmental or health problems. The use of PGRs in orchards is a very dynamic field. Scientists and chemical companies are continually looking for compounds that will improve fruit quality without unwanted side effects. In addition, new cultivars of fruits are continually being developed, and their cultivation and response to PGRs must be tested. Therefore, what is in use today may not be in use tomorrow. However, PGRs have a very important role to play in fruit development and maturation and will continue to be used for these purposes.

REFERENCES

Abeles, F.B., P.W. Morgan, and M.E. Saltveit (1992). *Ethylene in Plant Biology,* Second Edition. San Diego, CA: Academic Press.

Agusti, M., V. Almela, and J.L. Guardiola (1981). The regulation of fruit cropping in mandarins through the use of growth regulators. *Proceedings of the International Society of Citriculture* 1:216-220.

Arteca, R.N. (1996). Physiology of fruit set, growth, development, ripening, premature drop and abscission. In *Plant Growth Substances: Principles and Applications,* R.N. Arteca (Ed.). New York: Chapman and Hall, pp. 200-222.

Avery, G.S. Jr. and E.B. Johnson (1947). *Hormones and Horticulture: The Use of Special Chemicals in the Control of Plant Growth.* New York: McGraw-Hill Book Company.

Bangerth, F. (1978). The effect of a substituted amino acid on ethylene biosynthesis, respiration, ripening and preharvest drop of apple fruits. *Journal of the American Society of Horticultural Science* 103:401-404.

Ben Arie, R., H. Bazak, and A. Blumenfeld (1986). Gibberellin delays harvest and prolongs storage life of persimmon fruits. *Acta Horticulturae* 179:807-813.

Ben Arie, R. and S. Guelfat-Reich (1975). Early ripening of nectarines induced by succinic acid 2,2 dimethyl hydrazide (SADH) and 2-chloroethylphosphonic acid (CEPA) (in French). In *Facteurs et Regulation de la Maturation des Fruits.* Paris, France: Centre National de la Recherche Scientifique, pp. 24-29.

Ben Arie, R., Y. Roisman, Y. Zuthi, and A. Blumenfeld (1989). Gibberellic acid reduces sensitivity of persimmon fruits to ethylene. In *Biochemical and Physiological Aspects of Ethylene Production in Lower and Higher Plants,* H. Clijsters, H. De Profit (Eds.). Amsterdam, The Netherlands: Kluwer Academic Publishers, pp. 165-171.

Ben Arie, R., P. Sarig, Y. Cohen-Ahdut, Y. Zutkhi, L. Sonego, T. Kapulonov, and N. Lisker (1997). CPPU and GA_3 effects on pre- and post-harvest quality of seedless and seeded grapes. *Acta Horticulturae* 463:349-355.

Blanco, A. (1988). Control of shoot growth of peach and nectarine trees with paclobutrazol. *Journal of Horticultural Science* 63:201-207.

Blanco, A. (1990). Effects of paclobutrazol and ethephon on cropping and vegetative growth of 'Crimson Gold' nectarine trees. *Scientia Horticulturae* 42:65-73.

Blankenship, S.M., and E.C. Sisler (1993). Response of apples to diazocyclopentadiene inhibition of ethylene binding. *Postharvest Biology and Technology* 3:95-101.

Blommaert, K.L.J., A.N. Hanekom, and J. Steenkamp (1984). Cultivation guide to 'Barlinka' grapes. *Deciduous Fruit Grower* 34:8.

Blommaert, K.L.J., T. Theron, and J. Steenkamp (1975). Earlier and more uniform ripening of 'Santa Rosa' plums using ethephon. *Deciduous Fruit Grower* 25:267-271.

Bound, S.A., K.M. Jones, T.B. Koen, and M.J. Oakford (1991). The thinning effect of benzyladenine on red 'Fuju' apple trees. *Journal of Horticultural Science* 66:789-794.

Bound, S.A., K.M. Jones, T.B. Koen, M.J. Oakford, M.H. Barrett, and N.E. Stone (1991). The interaction of cytolin and NAA on cropping 'Red Delicious' apples. *Journal of Horticultural Science* 66:559-567.

Bradley, M.V. and J. Crane (1960). Gibberellin-induced inhibition of bud development in some species of *Prunus. Science* 131:825-826.

Byers, R.E. and D.H. Carbaugh (1991). Effect of chemical thinning sprays on apple fruit set. *HortTechnology* 1:41-48.

Christodoulou, A., R.J. Weaver, and R.M. Pool (1968). Relation of gibberellin treatment to fruitset, berry development and cluster compactness in *Vitis vinefera* grapes. *Procedings of the American Society of Horticultural Science* 92:226-230.

Church, R.M., L. Copas, and R.R. Williams (1984). Changes in fruit set, leaf size, and shoot growth or apple causes by some fungicides, insecticides, and a plant growth regulator. *Journal of Horticultural Science* 59:161-164.

Coggins, C.W. (1981). The influence of exogenous growth regulators on rind quality and internal quality of citrus fruits. *Proceedings of the International Society of Citriculture* 1:214-216.

Cohen, E., F.S. Latter, and R. Barkai-Golan (1965). The effect of NAA, 2,4,5-T and 2,4-D on the germination and development *in vitro* of fungi pathogenic to fruits. *Israel Journal of Agricultural Research* 15:41-47.

Curry, E.A. and K.M. Jones (1990). Controlling growth of 'Delicious' apple trees with low-dosage applications of uniconazole or BAS111. *Journal of Horticultural Science* 65:619-626.

Curry, E.A. and M.W. Williams (1986). Effect of paclobutrazol on fruit quality: Apple, pear, and cherry. *Acta Horticulturae* 179:743-753.

Davies, F.S. (1986) Growth regulator improvement of postharvest quality. In *Fresh Citrus Fruits,* W.F. Wardowski, S. Nagy, and W. Grierson (Eds.). Westport, CT: AVI Publishing Company, pp. 79-101.

Davis, T.D., G.L. Steffens, and N. Sankhla (1988). Triazole plant growth regulators. *Horticultural Review* 10:63-105.

El-Banna, G.I. and R.J. Weaver (1979). Effect of ethephon and gibberellin on maturation of ungirdled 'Thompson Seedless' grapes. *American Journal of Enology and Viticulture* 30:11-13.

Elfving, D.C. (1984). Factors affecting apple tree response to chemical branch-induction treatments. *Journal of the American Society of Horticultural Science* 109:476-481.

Elfving, D.C. (1989). N-(phenylmethyl)-1H-purine-6-amine (BA) as a chemical thinner for 'Idared' apple. *Acta Horticulturae* 239:357-362.

Elfving, D.C., C.L. Chu, E.C. Lougheed, and R.A. Cline (1987). Effects of daminozide and paclobutrazol treatments on fruit ripening, and storage behavior of 'McIntosh' apple. *Journal of the American Society of Horticultural Science* 112:910-915.

Elfving, D.C. and R.A. Cline (1993a). Benzyladenine and other chemicals for thinning 'Empire' apple trees. *Journal of the American Society of Horticultural Science* 118:593-598.

Elfving, D.C. and R.A. Cline (1993b). Cytokinin and ethephon affect crop load, shoot growth, and nutrient concentration of 'Empire' apple trees. *HortScience* 28:1011-1014.

Elfving, D.C. and E.C. Lougheed (1994). Storage responses of 'Empire' apples to benzyladenine and other chemical thinners. *Journal of the American Society of Horticultural Science.* 119:253-257.

Elfving, D.C., E.C. Lougheed, C.L. Chu, and R.A. Cline (1990). Effect of daminozide, paclobutrazol, and uniconazole treatments on 'McIntosh' apples at harvest and following storage. *Journal of the American Society of Horticultural Science* 115:750-756.

Elfving, D.C. and J.T.A. Proctor (1986). Long-term effects of paclobutrazol (Cultar) on apple-tree shoot growth, cropping, and fruit-leaf relations. *Acta Horticulturae* 179:473-480.

El-Khoreiby, A.M., C.R. Unrath, and L.J. Lehman (1990). Pactobutrazol spray timing influences apple tree growth. *HortScience* 25:310-312.

El-Otmani, M., A.A. M'Barek, and C.W. Coggins (1990). GA_3 and 2,4-D prolong on-tree storage of citrus in Morocco. *Scientia Horticulturae* 44:241-249.

El-Zeftawi, B.M. (1980a). Effects of gibberellic acid and cycocel on colouring and sizing of lemons. *Scientia Horticulturae* 12:177-181.

El-Zeftawi, B.M. (1980b). Regulating pre-harvest fruit drop and the duration of the harvest season of grapefruit with 2,4-D and GA. *Journal of Horticultural Science* 55:211-217.

Facteau, T.J. (1982). Levels of pectic substances and calcium in gibberellic acid-treated sweet cherry fruit. *Journal of the American Society of Horticultural Science* 107:148-151.

Facteau, T.J. and N.E. Chestnut (1991). Growth, fruiting, flowering, and fruit quality of sweet cherries treated with paclobutrazol. *HortScience* 26:276-278.

Facteau, T.J., K.E. Rowe, and N.E. Chestnut (1985). Response patterns of gibberellin acid-treated sweet cherry fruit at different soluble solids levels and leaf/fruit ratio. *Scientia Horticulturae* 27:257-262.

Fellman, C.D., P.E. Read, and M.A. Hosier (1987). Effects of thidiazuron and CPPU on meristem formation and shoot proliferation. *HortScience* 22:1197-1200.

Ferguson, L., F.S. Davies, M.A. Ismail, and T.A. Wheaton (1984). Growth regulator and low volume irrigation effects on grapefruit quality and fruit growth. *Scientia Horticulturae* 23:35-40.

Forshey, C.G. (1987). A review of chemical fruit thinning. *Proceedings of the Massachusetts Fruit Growers' Association* 93:68-73.

Forsyth, F.R., A.D. Crowe, H.J. Lightfoot, and G.L. Brown (1977). Effects of harvest date on the condition after storage of 'McIntosh' apples treated with ethephon and fenoprop. *Canadian Journal of Plant Science* 57:791-795.

Ga'ash, D. and S. Lavee (1982). The effect of growth regulators on maturation, quality and pre-harvest drop of stone fruits. *Acta Horticulturae* 34:449-455.

Garcia-Luis, A., A. Herrero-Villen, and J.L. Guardiola (1992). Effects of applications of gibberellic acid on late growth, maturation and pigmentation of the 'Clementine' mandarine. *Scientia Horticulturae* 49:71-82.

Gardner, F.E., P.C. Marth, and LP. Batjer (1940). Spraying with plant growth substances for control of the preharvest drop of apples. *Proceedings of the American Society of Horticultural Science* 37:415-428.

George, A.P., R.J. Nissen, J.A. Campbell, T. Rasmussen, and P. Allan (1995). Effects of paclobutrazol on growth and yield of low chill peaches in subtropical Australia. *Acta Horticulturae* 409:109-116.

Greany, P.D., R.E. McDonald, W.J. Schroeder, P.E. Shaw, M. Aluja, and A. Malavasi (1994). Use of gibberellic acid to reduce citrus fruit susceptibility to fruit flies. In *Bioregulators for Crop Protection and Pest Control*, P.A. Hedin (Ed.). American Chemical Society Symposium, 557:39-48.

Greene, D.W. (1986). Effect of paclobutrazol and analogs on growth, yield, and fruit quality, and storage potential of 'Delicious' apples. *Journal of the American Society of Horticultural Science* 111:328-332.

Greene, D.W. (1989). CPPU influences 'McIntosh' apple crop load and fruit characteristics. *HortScience* 24:94-96.

Greene, D.W. (1991). Reduced rates and multiple sprays of paclobutrazol control growth and improve fruit quality of 'Delicious' apples. *Journal of the American Society of Horticultural Science* 116:807-812.

Greene, D.W. and W.R. Autio (1989). Evaluation of benzyladenine as a chemical thinner on 'McIntosh' apples. *Journal of the American Society of Horticultural Science* 114:68-73.

Greene, D.W., W.R. Autio, and P. Miller (1990). Thinning activity of benzyladenine on several apple cultivars. *Journal of the American Society of Horticultural Science* 115:394-400.

Greene, D.W., W.J. Lord, and W.J. Bramlage (1977). Mid-summer applications of ethephon and daminozide on apples. II. Effect on 'Delicious'. *Journal of the American Society of Horticultural Science* 102:494-497.

Grierson, W., E. Cohen, and H. Kitagawa (1986). Degreening. In *Fresh Citrus Fruits*, W.F. Wardowski, S. Nagy, and W. Grierson (Eds.). Westport, CT: AVI Publishing Company, pp. 254-275.

Gross, J., H. Bazak, A. Blumenfeld, and R. Ben Arie (1984). Changes in chlorophyll and carotenoid pigments in the peel of 'Triumph' persimmon (*Diospyros kaki* L.) induced by pre-harvest gibberellin (GA_3) treatment. *Scientia Horticulturae* 24:305-314.

Guelfat-Reich, S. and Ben Arie, R. (1975). Maturation and ripening of 'Canino' apricot as affected by combined sprays of succinic acid 2,2-dimethylhydrazide (SADH) and 2,4,5 tri-chlorophenoxypropionic acid (2,4,5-TP). *Journal of the American Society of Horticultural Science* 100:517-519.

Gustafson, F.G. (1936). Inducement of fruit development by growth-promoting chemicals. *Proceedings of the National Academy of Sciences, USA* 22:628-636.

Harrell, D.C. and L.E. Williams (1987). The influence of girdling and gibberellic acid application at fruit set on 'Ruby Seedless' and 'Thompson Seedless' grapes. *American Journal of Enology and Viticulture* 38:83-88.

Hasegawa, K., N. Kuge, T. Mimura, and Y. Nakajima (1991). Effects of KT-30 and GA₃ on the fruit set and fruit growth of persimmon cvs. Saijo and Hiratanenashi. *Journal of the Japanese Society of Horticultural Science* 60:19-29.

Hislop, R.G. and R.J. Prokopy (1981). Integrated management of phytophagous mites in Massachusetts apple orchards. 2. Influence of pesticides on the predator *Amlyseius fallacis* under laboratory and field conditions. *Protective Ecology* 3:157-172.

Hull, J. Jr. and L.N. Lewis (1959). Response of one-year-old cherry and mature bearing cherry, peach and apple trees to gibberellin. *Proceedings of the American Society of Horticultural Science* 66:70-72.

Iwahori, S., S. Tominaga, and T. Yamasaki (1988). Stimulation of fruit growth of kiwifruit, *Actinidia chinensis* Planch., by N-(2-chloro-4-pyridyl)-N'-phenylurea, a diphenylurea derivative cytokinin. *Scientia Horticulturae* 35:109-115.

Johnson, D.S. (1994). Influence of time of flower and fruit thinning on the firmness of 'Cox's Orange Pippin' apples at harvest and after storage. *Journal of Horticultural Science* 69:197-203.

Jones, K.M., P. Jotic, T.B. Koen, S.B. Longley, and G. Adams (1988). Restructuring and cropping large 'Red Delicious' apple trees with paclobutrazol and daminozide. *Journal of Horticultural Science* 63:19-25.

Jones, K.M., T.B. Koen, and R.J. Meredith (1983). Thinning Golden Delicious apples using ethephon sprays. *Journal of Horticultural Science* 58:381-388.

Jones, K.M., T.B. Koen, M.J. Oakford, and S. Bound (1989). Thinning 'Red Fuji' apples with ethephon or NAA. *Journal of Horticultural Science* 64:527-532.

Knight, J.N. (1980). Fruit thinning of the apple cultivar Cox's Orange Pippin. *Journal of Horticultural Science* 55:267-273.

Knight, J.N., J.E. Spencer, N.E. Looney, and J.D. Lovell (1987). Chemical thinning of the apple cultivar Spartan. *Journal of Horticultural Science* 62:135-139.

Koen, T.B., K.M. Jones, and S.B. Longley (1988). Spray thinning strategies for 'Red Delicious' apple using naphthalene acetic acid and ethephon. *Journal of Horticultural Science* 63:31-35.

Kuraoka, T., K. Iwasaki, and T. Ishii (1977). Effects of GA₃ on puffing and levels of GA₃-like substances and ABA in the peel of 'Satsuma' mandarin (*Citrus unshiu* Marc.). *Journal of the American Society of Horticultural Science* 102:651-654.

Kurian, R.M. and C.P.A. Iyer (1993a). Chemical regulation of tree size in mango (*Mangifera indica* L.) cv. Alphonso. I. Effects of growth retardants on vegetative growth and tree vigour. *Journal of Horticultural Science* 68:349-354.

Kurian, R.M. and C.P.A. Iyer (1993b). Chemical regulation of tree size in mango (*Mangifera indica* L.) cv. Alphonso. II. Effects of growth retardants on flowering and fruit set. *Journal of Horticultural Science* 68:355-360.

Lavee, S. (1981). Control of grapevine growth and development, using ethylene releasing substances and some aspects of their action. *American Journal of Enology and Viticulture* 32:126-131.

Lima, J.E.O. and F.S. Davies (1984). Growth regulators, fruit drop, yield and quality of naval orange in Florida. *Journal of the American Society of Horticultural Science* 109:81-84.

Liu, S.H., C. Bussi, J. Hugard, and H. Clanet (1989). Critical period of flower bud induction in peach trees associated with shoot length and bud position (in German). *Gartenbauwissenschaft* 54:49-53.

Looney, N.E. (1972). Effects of succinic acid 2,2-dimethylhydrazide, 2-chloro-ethylphosphonic acid, and ethylene on respiration, ethylene production, and ripening of.'Redhaven' peaches. *Canadian Journal of Plant Science* 52:73-80.

Looney, N.E. (1975). Control of ripening in 'McIntosh' apples. II. Effect of growth regulators and CO_2 on fruit ripening, storage behavior, and shelf life. *Journal of the American Society of Horticultural Science* 100:332-336.

Looney, N.E. (1997). Hormones and horticulture. *HortScience* 32:1014-1018.

Looney, N.E. and P.D. Lidster (1980). Some growth regulator effects on fruit quality, mesocarp composition, and susceptibility to postharvest surface marking of sweet cherries. *Journal of the American Society of Horticultural Science* 105:130-134.

Looney, N.E. and J.E. McKellar (1987). Effect of foliar and soil surface applied paclobutrazol on vegetative growth and fruit quality of sweet cherries. *Journal of the American Society of Horticultural Science* 112:71-76.

Lurie, S., A. Ben Porat, Z. Lapsker, Y. Zuthi, Y. Greenblat, and R. Ben Arie (1997). Effect of spring applications of paclobutrazol and uniconazole on 'Red Rosa' plum fruit development and storage quality. *Journal of Horticultural Science* 72:93-99.

Lurie, S., S. Meir, and R. Ben Arie (1989). Preharvest ethephon sprays reduce superficial scald of 'Granny Smith' apples. *HortScience* 24:104-106.

Martin, G.C., F. Yoshikawa, and J.H. LaRue (1987). Effect of soil applications of paclobutrazol on vegetative growth, pruning time, flowering, yield, and quality of 'Flavorcrest' peach. *Journal of the American Society of Horticultural Science* 112:915-921.

McDonald, R.E., P.D. Greany, P.E. Shaw, and T.G. McCullum (1997). Preharvest applications of gibberellic acid delay senescence of Florida grapefruit. *Journal of Horticultural Science* 72:461-468.

Miliou, P.G. and E.M. Sfakiotakis (1986). Growth retardation activity of paclobutrazol on 'Delicious' apples. *Acta Horticulturae* 179:563-566.

Miller, S.S. (1988). Plant bioregulators in apple and pear culture. *Horticultural Review* 10:309-401.

Morris, J.R. and D.L. Cawthon (1982). Ethephon as a harvesting aid for 'Concord' grapes. *Arkansas Farm Research* 31:15.

Patterson, K.J., K.A. Mason, and K.A. Gould (1993). Effects of CPPU (N-(2-chloro-4-pyridyl)-N'-phenylurea) on fruit growth, maturity, and storage quality of kiwifruit. *New Zealand Journal of Crop and Horticultural Science* 21:253-261.

Patterson, W.K. and B.W. Zoecklein (1990). Vegetative and berry chemistry response to canopy manipulation and ethephon in 'Norton' grapes. *HortScience* 25:905-908.

Prive, J.P., D.C. Elfving, and J.T.A. Proctor (1989). Paclobutrazol, gibberellin, and cytokinin effects on growth, development, and histology of apple pedicels and fruit. *Journal of the American Society of Horticultural Science* 114:273-278.

Quinlan, J.D. and P.J. Richardson (1986). Uptake and translocation of paclobutrazol and implications for orchard use. *Acta Horticulturae* 179:443-451.

Rogers, B.L. and G.R. Williams (1977). Chemical thinning of spur-type 'Delicious' apple trees. *Virginia Fruit* 65:23-28.

Sansavini, G.R., R. Bonomo, A. Finotti, and V. Palara (1986). Foliar and soil applications of paclobutrazol on 'Gloster' apple. *Acta Horticulturae* 179:489-496.

Schneider, G.W. and J.V. Enzie (1944). Further studies on the effect of certain chemicals on the fruit set of apple. *Proceedings of the American Society of Horticultural Science* 45:63-68.

Sharples, R.O. (1973). Chemical control of growth and cropping—The influence of chemical growth regulators on fruit ripening and storage quality. *Scientia Horticulturae* 24:175-180.

Smart, R.E. (1982). Vine manipulation to improve wine grape quality. *Symposium of Proceedings of the Grape and Wine Centennial*, Davis, CA, pp. 362-375.

Southwick, F.W., I.E. Demoranville, and J.F. Anderson (1953). The influence of some growth regulating substances on preharvest drop, color and maturity of apples. *Proceedings of the American Society of Horticultural Science* 61:155-162.

Southwick, S.M., K.G. Weis, J.T. Yeager, and H. Zhou (1995). Controlling cropping in 'Loadel' cling peach using gibberellin: Effects on flower density, fruit distribution, fruit thinning, fruit firmness and yield. *Journal of the American Society of Horticultural Science* 120:1087-1095.

Southwick, S.M. and J.T. Yeager (1995). Use of gibberellin formulations for improved fruit firmness and chemical thinning of 'Patterson' apricot. *Acta Horticulturae* 384:425-429.

Southwick, S.M., J.T. Yeager, and H. Zhou (1995). Flowering and fruiting in 'Patterson' apricot in response to postharvest application of gibberellic acid. *Scientia Horticulturae* 60:267-277.

Steffens, G.L., F.W. Jacobs, and M.E. Engelhaupt (1993). Size, flowering and fruiting of maturing own-rooted 'Gala' apple trees treated with paclobutrazol sprays and trunk drenches. *Scientia Horticulturae* 56:13-21.

Tartarini, S., S. Sansavini, and M. Ventura (1993). CPPU control of fruit morphogenesis in apple. *Scientia Horticulturae* 53:273-279.

Veinbrants, N. (1975). Studies on the fruit thinning effect of NAA and some carbamates on 'Jonathan' and 'Delicious' apples. *Australian Journal of Experimental Agriculture and Animal Husbandry* 15:425-428.

Volz, R.K. and J.N. Knight (1986). The use of growth regulators to increase precocity in apple trees. *Journal of Horticultural Science* 61:181-189.

Wang, C.Y. and G.L. Steffens (1987). Postharvest responses of 'Spartan' apples to preharvest paclobutrazol treatment. *HortScience* 22:276-278.

Wang, C.Y., G.L. Steffens, and M. Faust (1986). Breaking bud dormancy in apple with a plant bioregulator, thiadizuron. *Phytochemistry* 25:311-317.

Watkins, C.B., E.W. Hewett, C. Bateup, A. Gunson, and C.M. Triggs (1989). Relationships between maturity and storage disorders in 'Cox's Orange Pippin' apples as influenced by preharvest calcium or ethephon sprays. *New Zealand Journal of Crop and Horticultural Science* 17:283-292.

Watkins, C.B., E. Stover, J.B. Halsey, and C.J. Torrice (1997). Retain—Experiences with 'McIntosh' and 'Jonagold' in New York. *Proceedings of Harvesting, Handling and Storage Workshop*, Cornell, NY: New York Agricultural Experiment Station, pp. 19-25.

Weaver, R.J. (1972). *Plant Growth Substances in Agriculture*. San Francisco, CA: W.H. Freeman and Company.

Weaver, R.J. (1975). Effect of growth retardant sprays on fruitfulness and cluster development of 'Thompson Seedless'. *American Journal of Enology and Viticulture* 26:47-49.

Weaver, R.J. and R.M. Pool (1965). Bloom spraying with gibberellin loosens clusters of 'Thompson Seedless' grapes. *California Agriculture* November:14-16.

Webster, A.D., J.D. Quinlan, and P.J. Richardson (1986). The influence of paclobutrazol on the growth and cropping of sweet cherry cultivars. I. The effect of annual soil treatments on the growth and cropping of cv. Early Rivers. *Journal of Horticultural Science* 61:471-478.

Whiley, A.W., J.B. Saranah, B.N. Wolstenholme, and T.S. Rasmussen (1991). Use of paclobutrazol sprays at mid-anthesis for increasing fruit size and yield of avocado (*Persea americana* Mill. cv. Hass). *Journal of Horticultural Science* 66:593-600.

Williams, M.W. and L.J. Edgerton (1981). Fruit thinning of apples and pears with chemicals. *Agricultural Information Bulletin* 289:1-9.

Wills, R.B.H. (1998). Enhancement of senescence in non-climacteric fruit and vegetables by low ethylene levels. *Acta Horticulturae* 463:349-355.

Zilkah, S., S. Lurie, Z. Lapsker, Y. Zuthi, I. Davis, Y. Yesselson, S. Antman, and R. Ben Arie (1997). The ripening and storage quality of nectarine fruits in response to preharvest application of gibberellic acid. *Journal of Horticultural Science* 72:355-362.

Zuluaga, E.M., J. Lumelli, and J.H. Christensen (1968). Influence of growth regulators on the characteristics of berries of *Vitis vinifera* L. *Phyton* 25:35-48.

Chapter 7

Role of PGRs in Citriculture

Kiyohide Kojima

INTRODUCTION

The production of seedless fruits has been a major objective. Thus, the mechanism of parthenocarpy has been studied. Auxins are most effective in inducing parthenocarpy in multiseeded fruits, whereas gibberellins (GAs) are effective in fruits with a few ovules (Naylor, 1984). Fruitlet growth and abscission, which are related to parthenocarpy, are regulated by several phytohormones (Schwabe and Mill, 1981). The peaks of abscisic acid (ABA) and indole-3-acetic acid (IAA) levels in fruitlets of 'Satsuma' mandarin were reported (Takahashi et al., 1975). It was suggested that more than one hormone may be involved in the fruit set (Garcia-Papi and Garcia-Martinez, 1984). Talon and colleagues (1992) suggested that the endogenous GA content in developing ovaries was the limiting factor controlling parthenocarpy. ABA plays an inhibitory role in growth (Zeevaart and Creelman, 1988), but its promotive effects have also been reported in fruit development (Kojima, Yamada, and Yamamoto, 1995; Ofosu-Anim, Kanayama, and Yamaki, 1996).

The author is grateful to Prof. N. Sakurai (Hiroshima University), Dr. A. Goto (Fruit Tree Research Station), and Dr. M. Yamamoto (Kagoshima University), for the use of facilities and valuable suggestions, and C. Kojima for technical assistance.

197

Abscission is controlled directly by phytohormones (Addicott, 1982). Goren (1993) suggested that the abscission of citrus fruits may be induced by ABA through ethylene synthesis. Auxins delay fruit abscission generally. The style of a flower was assumed to act as a sense organ for wilting (Gillissen, 1976) and showed the highest level of ABA in citrus (Goldschmidt, 1980; Harris and Dugger, 1986), but no direct causal relationship between the ABA level and pollination has been demonstrated. El-Otmani and colleagues (1995) have reviewed the published information on plant growth regulators (PGRs) of citrus.

This chapter provides information regarding ABA, IAA, and GA levels in both pollinated and parthenocarpic fruitlets and uniconazole effects on citrus to define possible relationships with fruit growth and abscission.

PROCEDURE FOR ABA, IAA, AND GA ANALYSES

Purification and Fractionation

Hormone fractionation was performed according to the following method (Kojima, Yamada, and Yamamoto, 1994; Kojima, Goto, and Yamada, 1995). The sample plus ^{3}H-ABA and ^{13}C-IAA was homogenized and filtered. The aqueous filtrate after evaporation was adjusted to pH 2.8 and filtered through membrane filters. The filtrate was partitioned against diethyl ether. The dried extract was fractionated with a high-performance liquid chromatography (HPLC) system. The effluents corresponding to the retention times of ABA and IAA were collected. The remaining effluent was collected for GA analysis. The aqueous phase after diethyl ether extraction was partitioned against ethyl acetate. The ethyl acetate layer was partitioned against 0.5 molar concentration (M) potassium hydrogen phosphate (K_2HPO_4). The aqueous phase was adjusted to pH 2.5 and was partitioned against ethyl acetate. The dried ethyl acetate layer and the effluent, except for the ABA and IAA fractions in the ether extract, were combined and purified by Sepralyte diethylaminopropyl (DEA).

Determination of ABA, IAA, and GA Levels

ABA content was determined by the method of Kojima and colleagues (Kojima, Kuraishi, Sakurai, Itou, and Tsurusaki, 1993; Kojima et al., 1995). The methylated ABA fraction was injected into a gas chromatograph using a capillary column equipped with a ^{63}Ni electron capture detector. A portion of methylated sample was injected into the HPLC system for collection of the methylated ABA. The radioactivity of the collected fraction was measured for the correction by recovery rates.

IAA content was determined by the methods of Cohen and colleagues (1986). The methylated IAA fraction was injected into a gas chromatograph using a capillary column equipped with a mass spectrometer.

GA content was bioassayed by the modified microdrop bioassay using dwarf rice seedling (Kojima, 1995; Kojima, Goto, and Yamada, 1995). The collected sample after drying was diluted successively three times with 50 percent acetone. The highest of all values over detectable levels obtained from gradually diluted samples was used as respective GA values.

ENDOGENOUS PHYTOHORMONES IN FLOWERS

Several twenty-year-old 'Hyuganatsu' trees [*Citrus tamurana* (Hort.) Ex. Tanaka] and 'Satsuma' mandarin trees (*Citrus unshiu* Marc.) growing in the same experimental orchard in the same year were used. 'Hyuganatsu' is a self-incompatible cultivar and has no potential for setting parthenocarpic fruits (Yamashita, 1978). 'Satsuma' mandarin is male-sterile and shows strong parthenocarpy. At least sixty buds and flowers from each were sampled and, in a chilled room (10°C), immediately separated into parts (see the upper insert in Figure 7.1).

ABA Levels in Flowers

In 'Hyuganatsu' and 'Satsuma' mandarin, the highest ABA concentrations were observed in styles (see the upper insert in Figure 7.1). The same tendency was observed in 'Shamouti' orange (Goldschmidt, 1980). ABA levels of the style in 'Washington navel' orange increased without pollination (Harris and Dugger, 1986). In

FIGURE 7.1. Concentration of ABA, IAA, and GAs in Parts of Bud (Five Days Before Flowering) and Flower (at Flowering) in Citrus

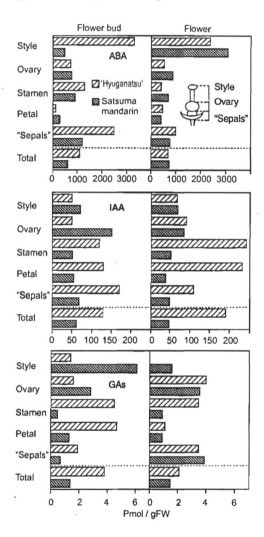

Source: These values were obtained from Kojima, 1996, pp. 267, 269; Kojima et al., 1996, pp. 239-240; and Kojima, 1997, pp. 273-274.

Note: GAs are expressed as GA_3 equivalents detected by rice bioassay. "Sepals" included floral disks, sepals, and receptacles.

'Satsuma' mandarin, ABA concentration of the style increased six-fold from the bud to flowering stage, confirming its status as a parthenocarpic citrus (see the upper insert in Figure 7.1). Moreover, pollination led to a threefold increase in ABA concentration in the style of 'Hyuganatsu' 8 days after flowering (DAF) (non-pollination, 3,900; pollination, 11,000 picomoles [pmol]/g fresh weight) (Kojima, 1996). Thus, the increase in the ABA concentration in the style by pollination may be the signal for wilting, and the style may be a sense organ.

In both cultivars, ABA concentrations in the stamens decreased at flowering, while those in petals increased (see the upper insert in Figure 7.1), confirming the results of Goldschmidt (1980). ABA concentrations in the stamens of 'Hyuganatsu' decreased by less than one-fourth. In tomato, ABA concentrations in the stamens also decreased at flowering (Kojima, Kuraishi, Sakurai, and Fusao, 1993). These decreases of ABA concentrations in stamens at flowering may be related to pollen maturation.

IAA Levels in Flowers

'Satsuma' mandarin showed lower concentrations of IAA in the stamens, petals, and "sepals" than 'Hyuganatsu' in both bud and flower stages (see the middle insert in Figure 7.1). In the stamens, 'Satsuma' mandarin, which forms aborted anthers, did not show any change in IAA concentrations from bud to flower stage, while 'Hyuganatsu', which forms normal pollen, doubled IAA concentration. In the tomato flower, which forms normal pollen, the IAA concentration of the stamens increased by more than four times during the same period (Kojima, Sakurai, and Tsurasaki, 1994). Thus, the rapid increase in the IAA concentration in 'Hyuganatsu' might be related to maturation of the pollen grains.

GA Levels in Flowers

'Hyuganatsu' had higher concentrations of GAs in the stamens than 'Satsuma' mandarin in both bud and flower stages (see the lower insert in Figure 7.1), suggesting that gibberellin might be related to the presence of normal pollen. GA concentrations in styles decreased considerably in both cultivars, but its physiological significance is not understood. The GA concentrations in ovaries and "sepals" increased (see the following fruitlet section).

CHANGES IN PHYTOHORMONE
LEVELS IN FRUITLETS

The plant materials and sampling methods were the same as in the case of flowers. Fruitlets with attachments were immediately separated into fruitlets and "sepals," which included floral disks, sepals, and receptacles (see Figure 7.2-I-e).

Abscission and Growth of Fruitlet

Early development can be divided into three phases in most fruits (Gillaspy, Ben-David, and Gruissem, 1993): phase I, period of ovary development, fertilization, and fruit set; phase II, period of cell division, seed formation, and early embryo development; and phase III, period of cell expansion and embryo maturation. In citrus, cell division occurs in phases I and II (Bain, 1958). Cell division lessens temporarily in phase I and may be reactivated by phytohormones from fertilized ovules (Pharis and King, 1985). Phase I corresponds to the period from 0 to about 10 DAF, judging from the observation of pollen tube growth (Yamashita, 1978). Phase II corresponds to the period from 10 to about 30 DAF, judging from the observation of cell division (Bain, 1958).

The pollinated fruitlets of 'Hyuganatsu' stopped falling after phase I, and nonpollinated fruitlets fell completely by 42 DAF (see Figure 7.2-I-a). The fruitlets of 'Satsuma' mandarin continued to drop at a lower rate because only flower buds with leaves were used for the experiment (see Figure 7.2-II-a).

The fresh weight of the pollinated fruitlets increased linearly on a logarithmic scale after flowering, whereas that of nonpollinated fruitlets increased at a lower rate (see Figure 7.2-I-b). "Sepals" after pollination showed slightly heavier fresh weights than those after nonpollination. In parthenocarpic mandarin, the fresh weight of fruitlets continued to increase after bud stage, whereas that of "sepals" increased slightly (see Figure 7.2-II-b).

FIGURE 7.2. Change in Abscission, Fresh Weight, and Concentrations of Phytohormones in Ovaries/Fruitlets, and "Sepals" of 'Hyuganatsu' (I) and 'Satsuma' Mandarin (II)

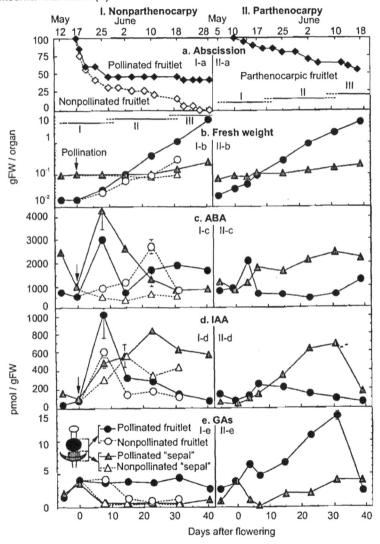

Source: These values were obtained from Kojima, 1996, p. 269; Kojima et al., 1996, p. 240; and Kojima, 1997, pp. 275-276.

Note: The upper horizontal axis shows sampling.

ABA Levels

ABA has been suggested to play an inhibitory role in growth (Zeevaart and Creelman, 1988), but promotive effects of ABA in sink tissues have also been reported (Kojima, Yamada, and Yamamoto, 1995; Ofosu-Anim, Kanayama, and Yamaki, 1996). As to the positive effects of ABA in sink organs, there are two hypotheses: (1) ABA functions as a promoter of sink activity (Brenner, Schreiber, and Jones, 1989), and (2) ABA induces stomatal closure and depresses photosynthesis in sources. Thus, sinks may accumulate ABA and increase the ability of assimilate production in sources (Brenner and Cheikh, 1995). The pollinated fruitlet showed a peak of ABA concentration in phase I (see Figure 7.2-I-c), and the parthenocarpic fruitlet also showed a peak (see Figure 7.2-II-c). Thus, if the previous hypotheses are valid, ABA in fruitlets may affect fruit set positively in both setting types in phase I.

In "sepals," changing patterns of ABA concentrations differed among pollinated, nonpollinated, and parthenocarpic fruitlets (see Figure 7.2-I, II-c). "Sepals" after pollination showed the highest peak of ABA concentration in phases I and II. "Sepals" after parthenocarpy contained higher concentrations than after nonpollination in phases II and III. Thus, that "sepals" in both setting types contained higher concentrations of ABA than those in nonsetting types suggests that ABA in "sepals" plays a beneficial role in fruitlet growth and/or retention.

IAA Levels

The pollinated fruitlets exhibited the highest peak in IAA concentration in phase I throughout the experimental period, and nonpollinated fruitlets also showed a peak (see Figure 7.2-I-d). Parthenocarpic fruitlets showed just a small peak in IAA concentration in phase I (see Figure 7.2-II-d). The application of synthetic auxins directly stimulated the fruitlet growth of mandarin (Guardiola and Lazaro, 1987). Thus, in the nonparthenocarpic cultivar, the IAA increase in phase I may promote growth of the ovary/fruitlet, and in the parthenocarpic cultivar, phytohormones other than IAA may influence growth.

GA Levels

Ben-Cheikh and colleagues (1997) proposed that pollination induces GA increases within fruitlets in seeded varieties of citrus. The results from 'Hyuganatsu' support this proposal (see Figure 7.2-I-e). Parthenocarpic fruitlets had higher concentrations of GAs than pollinated fruitlets (see Figure 7.2-I, II-e), confirming the results in grapes (Iwahori, Weaver, and Pool, 1968) and pears (Gil, Martin, and Griggs, 1972). The reproductive organ of the parthenocarpic mutant of *Citrus* contained slightly higher levels of GA_1, GA_2, and GA_4 than the pollinated cultivar (Talon, Zacarias, and Primo-Millo, 1992). My results supported the suggestion that the endogenous GA level in the developing ovaries in the parthenocarpic cultivars was the limiting factor controlling the fruit set (Talon, Zacarias, and Primo-Millo, 1992).

Both pollinated and parthenocarpic fruitlets showed higher concentrations of GAs than nonsetting fruitlets around phase II (see Figure 7.2-I, II-e). The localized GA application on *Citrus* pistils promoted the mobilization of assimilates into young ovaries (Powell and Krezdorn, 1977). Thus, the higher concentrations of endogenous GAs recorded around phase II may play a major role in the assimilate accumulation to ensure the fruit set. "Sepals" showed no difference between the GA concentrations during pollination and in the absence of pollination (see Figure 7.2-I-e).

ABSCISSION AND PHYTOHORMONES
OF FRUITLETS

It is thought that abscission is controlled directly by phytohormones (Addicott, 1982). ABA was originally discovered as an abscission-promoting hormone, and its promotive effect has been documented in several plants. In *Citrus* fruits, abscission may be induced by the following process (Goren, 1993): (1) ABA induces ethylene synthesis, (2) which increases the activity and production of hydrolytic enzymes, and (3) the enhanced enzymes degrade the cell wall in the abscission zone, (4) which results in fruit separation. On the other hand, auxins generally delay fruit abscission. Takahashi and colleagues (1975) suggested that a sharp decrease in auxin, as well as an ABA increase, might be important in fruitlet abscission of *Citrus*.

Effects of GA Biosynthesis Inhibitor

Uniconazole acts as an inhibitor of GA biosynthesis (Izumi et al., 1985). Uniconazole markedly accelerated fruitlet abscission of 'Satsuma' mandarin but only slightly decreased its growth (Kojima, Goto, and Nakashima, 1996). Paclobutrazol, an inhibitor of GA biosynthesis, also increased *Citrus* fruit abscission (Talon, Zacarias, and Primo-Millo, 1992). Uniconazole decreased GA concentration by less than one-tenth seven days after treatment, confirming that uniconazole inhibits GA biosynthesis in the *Citrus* fruitlet (see Table 7.1). Talon and colleagues (1992) suggested that GA may control the fruit development of parthenocarpic *Citrus* and implied the occurrence of a threshold for GA levels. Thus, fruitlet abscission may be due to the reduced GA levels in fruitlets. ABA concentrations in fruitlets treated with uniconazole were about fourfold higher than those in the control, both seven and twenty-one days after treatment (see Table 7.1). The paclobutrazol treatment also induced a threefold increase in the ABA level in 'Satsuma' mandarin fruits (Zacarias et al.,

TABLE 7.1. Concentrations of ABA, IAA, and GAs in 'Mandarin' Fruitlets Treated with Uniconazole

Treatment	ABA[a]	IAA[a] pmol / gFW	GAs[y]
7 days after treatment			
Control	640 ± 20	180 ± 1	5.5
Uniconazole	2300 ± 32	110 ± 7	0.3
21 days after treatment			
Control	1500 ± 37	51 ± 2	2.6
Uniconazole	5900 ± 96	78 ± 2	1.8

Source: These values were obtained from Kojima, Goto, and Nakashima, 1996, p. 902.

[a] Concentrations of ABA and IAA (three measurements) are mean ± SE.
[y] pmol GA_3 eq. / gFW.

1995). On the other hand, exogenous GAs inhibited ABA increase and abscission of *Citrus* fruit. Thus, endogenous ABA in the fruitlet that was increased by uniconazole may also promote the abscission of mandarin fruitlets by the sequential mechanism suggested by Goren (1993).

Endogenous Hormone Levels

In nonparthenocarpic 'Hyuganatsu', pollinated fruitlets showed a lesser extent of fruitlet abscission than nonpollinated fruitlets (see Figure 7.2-I-a). Pollinated fruitlets contained higher concentrations of ABA than nonpollinated fruitlets in phase I (see Figure 7.2-I-c). In addition, "sepals" of pollinated fruitlets contained much higher concentrations of ABA in phases I and II. These results did not support the hypothesis that endogenous ABA may promote fruitlet abscission in phases I and II.

"Sepals" after both pollination and parthenocarpy contained higher concentrations of IAA than fruitlets around phase II (see Figure 7.2-I, II-d), although I had assumed that they would exhibit lower concentrations of IAA in all phases. Auxin inhibited the rise in polygalacturonase and cellulase activity in abscission zones and delayed fruit abscission (Huberman and Goren, 1979). Iwahori and colleagues (1990) reported that IAA applied via peduncle markedly retarded abscission between a fruitlet and "sepal" of a Citrus explant. Thus, IAA from "sepals" may inhibit hydrolase activity and hinder fruitlet abscission.

REFERENCES

Addicott, F.T. (1982). Physiology. In *Abscission.* Berkeley and Los Angeles, CA: University of California Press, pp. 97-152.

Bain, J.M. (1958). Morphological, anatomical, and physiological changes in the developing fruit of the Valencia orange, *Citrus sinensis* (L.) Osbeck. *Australian Journal of Botany* 6:1-24.

Ben-Cheikh, W., J. Perez-Botella, F.R. Tadeo, M. Talon, and E. Primo-Millo (1997). Pollination increases gibberellin levels in developing ovaries of seeded varieties of citrus. *Physiologia Plantarum* 114:557-564.

Brenner, M.L. and N. Cheikh (1995). The role of hormones in photosynthate partitioning and seed filling. In *Plant Hormones: Physiology, Biochemistry and Molecular Biology,* Second Edition, P.J. Davies (Ed.). Dordrecht, Netherlands: Kluwer Academic Publishers, pp. 649-670.

Brenner, M.L., B.M.N. Schreiber, and R.J. Jones (1989). Hormonal control of assimilate partitioning: Regulation in the sink. *Acta Horticulturae* 239:141-148.

Cohen, J.D., B.G. Baldi, and J.P. Slovin (1986). [13]C-[benzene ring]-indole-3-acetic acid. A new internal standards for quantitative mass spectral analysis of indole-3-acetic acid in plants. *Plant Physiology* 80:14-19.

El-Otmani, M., C.J. Lovatt, C.W. Coggins Jr., and M. Agusti (1995). Plant growth regulators in citriculture: Factors regulating endogenous levels in citrus tissues. *Critical Review of Plant Science* 14:367-412.

Garcia-Papi, M.A. and J.L. Garcia-Martinez (1984). Endogenous plant growth substances content in young fruits of seeded and seedless Clementine mandarin as related to fruit set and development. *Scientia Horticulturae* 22:265-274.

Gil, G.F., G.C. Martin, and W.H. Griggs (1972). Fruit set and development in the pear: Extractable endogenous hormones in parthenocarpic and seeded fruits. *Journal of American Society for Horticultural Science* 97:731-735.

Gillaspy, G., H. Ben-David, and W. Gruissem (1993). Fruits: A developmental perspective. *Plant Cell* 5:1439-1451.

Gillissen, L.J.W. (1976). The role of the style as a sense-organ in relation to wilting of the flower. *Planta* 131:201-202.

Goldschmidt, E.E. (1980). Abscisic acid in citrus flower organs as related to floral development and function. *Plant and Cell Physiology* 21:193-195.

Goren, R. (1993). Anatomical, physiological, and hormonal aspects of abscission in citrus. *Horticultural Review* 15:145-182.

Guardiola, J.L. and E. Lazaro (1987). The effect of synthetic auxins on fruit growth and anatomical development in Satsuma mandarin. *Scientia Horticulturae* 31:119-130.

Harris, M.J. and W.M. Dugger (1986). Levels of free and conjugated abscisic acid in developing floral organs of the navel orange [*Citrus sinensis* (L.) Osbeck cv. Washington]. *Plant Physiology* 82:1164-1166.

Huberman, M. and R. Goren (1979). Exo- and endo-cellular cellulase and polygalacturonase in abscission zones of developing orange fruit. *Physiologia Plantarum* 45:189-196.

Iwahori, S., S. Tominaga, and S. Higuchi (1990). Retardation of abscission of citrus leaf and fruitlet explants by brassinolide. *Plant Growth Regulation* 9:119-125.

Iwahori, S., R.J. Weaver, and R.M. Pool (1968). Gibberellin-like activity in berries of seeded and seedless tokay grapes. *Plant Physiology* 43:333-337.

Izumi, K., K. Kamiya, A. Sakurai, H. Oshio, and N. Takahashi (1985). Studies of sites of action of a new plant growth retardant (E)-1-(4-chlorophenyl)-4,4-dimethyl-2-(1, 2, 4-triazol-1-ly)-1-pention-3-ol(S3307) and comparative effects of its stereoisomers in a cell-free system from cucurbita maxima. *Plant and Cell Physiology* 26:821-827.

Kojima, K. (1995). Simultaneous measurement of ABA, IAA and GAs in citrus— Role of ABA in relation to sink ability. *Japan Agricultural Research Quarterly* 29:179-185.

Kojima, K. (1996). Changes of abscisic acid, indole-3-acetic acid and gibberellin-like substances in the flowers and developing fruitlets of citrus cultivar 'Hyuganatsu'. *Scientia Horticulturae* 65:901-902.

Kojima, K. (1997). Changes of ABA, IAA and GAs levels in reproductive organs of citrus. *Japan Agricultural Research Quarterly* 31:273-282.

Kojima, K., A. Goto, and S. Nakashima (1996). Effects of uniconazole-P on abscission and endogenous ABA, IAA and GA-like substances levels of Satsuma mandarin fruitlet. *Bioscience, Biotechnology, and Biochemistry* 60:901-902.

Kojima, K., A. Goto, and Y. Yamada (1995). Simultaneous measurement for ABA, IAA and GAs in citrus fruits. *Bulletin of Fruit Tree Research Station* 27:1-10.

Kojima, K., S. Kuraishi, N. Sakurai, and K. Fusao (1993b). Distribution of abscisic acid in different parts of the reproductive organs of tomato. *Scientia Horticulturae* 56:23-30.

Kojima, K., S. Kuraishi, N. Sakurai, T. Itou, and K. Tsurusaki (1993). Spatial distribution of abscisic acid and 2-trans-abscisic acid in spears, buds, rhizomes and roots of asparagus (*Asparagus officinalis* L.). *Scientia Horticulturae* 54:177-189.

Kojima, K., N. Sakurai, and K. Tsurusaki (1994). Distribution of IAA within flower and fruit of tomato. *HortScience* 29:1200.

Kojima, K., T. Takahara, T. Ogata, and N. Muramatsu (1995). Relationships between growth properties and endogenous ABA, IAA and GA of citrus varieties for rootstock (in Japanese with English summary). *Journal of the Japanese Society for Horticultural Science* 63:753-760.

Kojima, K., Y. Yamada, and M. Yamamoto (1994). Distribution of ABA and IAA within a developing valencia orange fruit and its parts (in Japanese). *Journal of the Japanese Society for Horticultural Science* 63:335-339.

Kojima, K., Y. Yamada, and M. Yamamoto (1995). Effects of abscisic acid injection on sugar and organic acid contents of citrus fruit. *Journal of the Japanese Society for Horticultural Science* 64:17-21.

Kojima, K., M. Yamamoto, A. Goto, and R. Matsumoto (1996). Changes of ABA, IAA and GAs contents in reproductive organs of Satsuma mandarin (in Japanese). *Journal of the Japanese Society for Horticultural Science* 65:237-243.

Naylor, A.W. (1984). Functions of hormones at the organ level of organization. In *Hormonal Regulation of Development. II. Encyclopedia of Plant Physiology*. New series, Volume 10. T.K. Scott (Ed.). Berlin/Heidelberg: Springer-Verlag, pp. 195-200.

Ofosu-Anim, J., Y. Kanayama, and S. Yamaki (1996). Sugar uptake into strawberry fruit is stimulated by abscisic acid and indoleacetic acid. *Plant Physiology* 97:169-174.

Pharis, R.P. and R.W. King (1985). Gibberellins and reproductive development in seed plants. *Annual Review of Plant Physiology* 36:517-568.

Powell, A.A. and A.H. Krezdorn (1977). Influence of fruit-setting treatment on translocation of ^{14}C-metabolites in citrus during flowering and fruiting. *Journal of American Society for Horticultural Science* 102:709-714.

Schwabe, W.W. and J.J. Mill (1981). Hormones and parthenocarpic fruit set. *Horticultural Abstract* 51:661-698.

Takahashi, N., I. Yamaguchi, T. Kono, M. Igoshi, K. Hirose, and K. Suzuki (1975). Characterization of plant growth substances in *Citrus unshiu* and their changes in fruit development. *Plant and Cell Physiology* 16:1101-1111.

Talon, M., L. Zacarias, and E. Primo-Millo (1992). Gibberellins and parthenocarpic ability in developing ovaries of seedless mandarins. *Plant Physiology* 99:1575-1581.

Yamashita, K. (1978). Studies on self-incompatibility of Hyuganatsu, *Citrus tamurana* Hort. ex Tanaka (in Japanese). *Journal of the Japanese Society for Horticultural Science* 47:188-194.

Zacarias, L., M. Talon, W. Ben-Cheikh, M.T. Lafuente, and E. Primo-Millo (1995). Abscisic acid increases in nongrowing and paclobutrazol-treated fruits of seedless mandarins. *Physiologia Plantarum* 95:613-619.

Zeevaart, J.A.D. and R.A. Creelman (1988). Metabolism and physiology of abscisic acid. *Annual Review of Plant Physiology* 39:439-473.

Chapter 8

Reducing Fruit Drop
in Lychee with PGR Sprays

Raphael A. Stern
Shmuel Gazit

INTRODUCTION

The lychee, which belongs to the Sapindaceae family (Tindall, 1994), produces one of the world's finest fruits (Galan-Sauco and Menini, 1989). It originated in China, where it has been cultivated for over 3,000 years, and has become widely distributed in the tropics and subtropics (Knight, 1980; Joubert, 1986; Galan-Sauco and Menini, 1989). It has not, however, become a major crop, mainly because of the problem of low and irregular yields (Knight, 1980; Samson, 1980; Menzel, 1983, 1984; Batten, 1986; Joubert, 1986; Galan-Sauco and Menini, 1989; Stern, 1992; Stern and Gazit, 1996, 1997, 1998; Stern, Adato, et al., 1993; Stern, Gazit, et al., 1993; Stern et al., 1995, 1996, 1997, 1998; Stern, Nadler, and Gazit, 1997). The fruit is a tubercled drupe, oval to ovoid in shape, about 3 to 3.5 centimeters (cm), in diameter and 3 to 4 cm long. It has a rough, brittle, and red husk. The flesh is juicy, white, and translucent. The single seed is usually large, but occasionally small and shrunken or abortive (Joubert, 1986; Galan-Souco and Menini, 1989; Tindall, 1994).

The poor yield can be partly attributed to low flowering intensity, especially after inadequate chilling, which induces vegetative dormancy prior to floral initiation (Menzel, 1983, 1984; Chaikiattiyos, Menzel, and Rasmussen, 1995). However, even after profuse flowering—which can be induced by autumnal water stress (Stern, Adato, et al., 1993; Stern et al., 1998)—the yield is usually inadequate, mainly as a result of massive fruit drop during the early period of fruit development (Menzel, 1983, 1984; Joubert, 1986; Yuan and Huang, 1988; Galan-Sauco and Menini, 1989; Stern et al., 1995; Stern, Nadler, and Gazit, 1997; Stern and Gazit, 1997, 1999).

This chapter reviews the literature concerning fruit growth and development, the abscission rate, and the use of plant growth regulators (PGRs) to reduce fruitlet abscission and increase yields.

FRUIT GROWTH AND DEVELOPMENT

Khan (1929) was the first to publish a description of lychee fruit growth. Generally, only one lobe of the bilobed ovaries develops. The fleshy, edible part of the fruit, called the aril, develops over the testa. Stern and colleagues (1995), who worked on 'Mauritius', a cultivar identical to the Chinese 'Tai So' and the Thai 'Hong Huey' (Menzel and Simpson, 1990), described three stages of fruit growth (see Figure 8.1). The first stage was completed in five weeks; during this stage, the main increment in weight is due to pericarp growth. By the end of this stage, the fruit weighs about 2 grams (g), and the embryo has two tiny developed cotelydons, visible to the naked eye when the fruit is cut lengthways (heart stage). The aril is only a negligible portion of the fruit at this stage.

The second stage of growth takes place at five to seven weeks after fruit set. Seed growth is completed, and the endosperm disappears. By the end of this phase, the embryo fills the seed cavity with well-developed cotyledons, the seed coat has hardened, and the aril is just commencing its rapid growth. The fruitlet weighs about 7 g.

The third stage takes place seven to thirteen weeks after fruit set and is characterized by the rapid growth of the aril. The weight of the mature fruit ranges from 20 to 28 g, of which the pericarp and seed comprise 10 percent each, and the aril 80 percent of the total.

The findings of Stern and colleagues (1995) are consistent with previous investigations of other lychee cultivars (Joubert, 1967; Kanwar, Nijar, and Rajput, 1972; Pivovaro, 1974; Guar and Bajpai, 1978; Huang and Xu, 1983; Chaitrakulsub, Chaidate, and Gemma, 1988).

FIGURE 8.1. Cumulative Growth of 'Mauritius' Lychee Fruit, Seed, Pericarp, and Aril

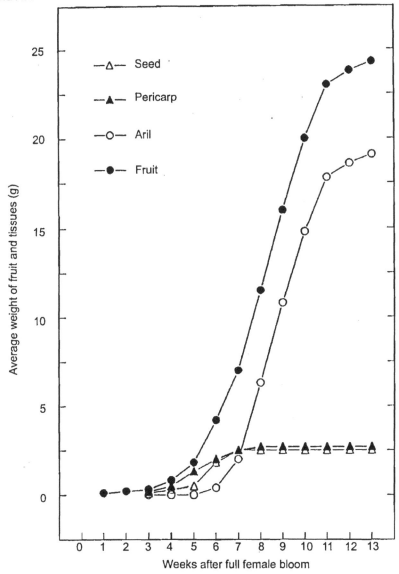

Source: Stern et al., 1995, pp. 65-70.

Note: Each value is an average of twelve replicates.

FRUITLET ABSCISSION PATTERN

The lychee tree produces a large number of inflorescences with three flower types: male, female, and pseudohermaphrodite (Joubert, 1986). Under normal conditions, each inflorescence carries 100 to 250 female flowers, of which only a very small percentage develop into mature fruit, due to massive flower and fruitlet abscission (Hayes, 1945; Mustard, Liu, and Nelson, 1953; Chandler, 1958; Mustard, 1960; Prasad and Jauhari, 1963; Chadha and Rajpoot, 1969; Hoda, Syamal, and Chhonkar, 1973; Misra, Nauriyal, and Awasthi, 1973; Pivovaro, 1974; Veera and Das, 1974; Singh and Lal, 1980; Yuan and Huang, 1988; McConchie and Batten, 1989; Stern et al., 1995; Stern, Nadler, and Gazit, 1997; Stern and Gazit, 1997, 1999). Most of the flowers and fruitlets abscise during the first month after pollination (Mustard, Liu, and Nelson, 1953; Joubert, 1986; Stern et al., 1995; Stern, Nadler, and Gazit, 1997; Stern and Gazit, 1999). It was found that most of the female flowers are abnormal, in that they lack an embryo sac, egg apparatus, and/or polar nuclei (Stern et al., 1996, 1997). Thus, most of the flowers are not fertilized, due to pollination or fertilization failure, and hence cannot set normal fruit (Joubert, 1986; McConchie and Batten, 1989; Stern and Gazit, 1996, 1998). The percentage of fruit drop during the developmental period varies according to locality, cultivar, and environmental and cultural conditions. In some cases, all the fruit of a given panicle abscise before harvest.

In Israel, 'Mauritius' retained almost 100 percent more fruit per panicle than 'Floridian', which may be identical with 'Brewster' and the Chinese 'Chen Zi' (Groff, 1948). This phenomenon was associated with higher production in 'Mauritius'. Similarly, but to an even more marked extent, 'Mauritius' outproduced 'Kaimana' (Stern and Gazit, 1999), and 'Calcutta' outproduced 'Early Seedless' in India (Singh and Lal, 1980). Stern and colleagues (Stern et al., 1995; Stern, Nadler, and Gazit, 1997; Stern and Gazit, 1999) noted two distinct abscission periods, with minor variations in the three cultivars examined. The first period lasts about four weeks, at the end of which 5 to 15 percent of the female flowers survive and develop into small fruitlets. One or two weeks later, a second wave of abscission begins, lasting about two weeks. All fruitlets that abscise at this stage contain a well-developed seed coat, and weigh 2 to 6 g. When the embryo has reached full size, the rate of abscission subsides (Stern et al., 1995). During this period, about half the remaining 'Mauritius' and 'Floridian' fruitlets abscise. Hence, only 1 to 5 percent of the initial female flowers finally develop into mature fruit.

Rapid embryo growth was found to coincide with the second abscission period (see Figure 8.2), a relationship that was observed consistently in several seasons (Stern et al., 1995).

REDUCTION OF FRUITLET ABSCISSION WITH PGR SPRAYS

Abscission of young fruit is a common phenomenon (Leopold and Kriedman, 1975; Addicott, 1982). In a number of species, including citrus, apples, pears, and peaches, it can be reduced or prevented by the application of an auxin (Leopold, 1958; Weaver, 1972; Arteca, 1996). Evidence suggests that auxins inhibit the action of the hydrolytic enzymes polygalacturonase and cellulase, which are responsible for the degradation of the cell wall and the middle lamella in the abscission zone (Goren, 1993). Ethylene promotes the synthesis and activity of these enzymes, whereas auxins delay these processes (Goren and Huberman, 1976). However, the specific relationship between auxin production in developing fruit and abscission in lychee has never been investigated. Liu (1986) found that the endogenous indoleacetic acid (IAA) content in lychee fruitlets rose steeply during the first three weeks of fruit development, from 150 to 850 micrograms per gram ($\mu g \cdot g^{-1}$) fresh weight, but decreased after the onset of rapid embryo development, four to five weeks after fertilization, falling to 300 $\mu g \cdot g^{-1}$.

Many studies have been conducted to examine the possibility of reducing abscission in lychee by the application of synthetic auxins. The main auxins tried were α-naphthaleneacetic acid (NAA), 2,4-dichlorophenoxyacetic acid (2,4-D), 2,4,5-trichlorophenoxyacetic acid (2,4,5-T), and 2,4,5-trichlorophenoxypropionic acid (2,4,5-TP). A number of these studies produced positive results (Prasad and Jauhari, 1963; Khan, Misra, and Srivastava, 1976; Yuan and Huang, 1991), whereas others failed to reduce abscission (Hoda, Syamal, and Chhonkar, 1973; Misra, Nauriyal, and Awasthi, 1973; Veera and Das, 1974; Singh and Lal, 1980; Verma, Jain, and Dass, 1981) or even increased abscission (Shoan and Dhillon, 1981). In Israel, lychee fruit abscission was totally inhibited after application of Tipimon (a liquid solution

FIGURE 8.2. Patterns of 'Mauritius' Lychee Fruitlet Abscission and Seed Growth

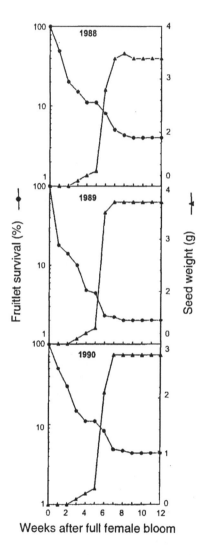

Source: Stern et al., 1995, pp. 67-70.

Note: Initial number of female flowers per inflorescence was about 160. Each value is an average of twelve replicates obtained in 1988, 1989, and 1990.

manufactured by Tapazol, containing 6.8 percent 2,4,5-TP, formulated as triethanolamine salt) 1 to 3 weeks after fertilization. However, most of the resulting mature fruits were seedless, very small, and had no market value (Pivovaro, 1974). In another trial, Tipimon failed to increase yield when applied 7 to 8 weeks after fertilization (Shalem-Galon, 1980).

At the beginning of the 1990s, it was found that fruitlet drop in 'Mauritius' was reduced, and the yield significantly increased, by spraying with 100 parts per million (ppm) 2,4,5-TP (0.15 percent Tipimon) at the stage when fruitlets weigh about 2 g (Stern et al., 1995). At this stage, the embryo begins its rapid growth, followed by the second wave of drop a few days later. The greater retention of fruit is apparently the result of a rise in the auxin content of the fruitlet through absorption. The increase in fruit with shriveled seeds indicates that the treatment reduced the abscission of fruitlets with degenerate or aborted embryos (Stern, Nadler, and Gazit, 1997).

The reported result that 2,4,5-TP decreased abscission in these experiments (Stern, 1992; Stern et al., 1995; Stern, Nadler, and Gazit, 1997), while NAA and 2,4-D did not, is in agreement with previous reports from Israel (Pivovaro, 1974; Shalem-Galon, 1980), but contradicts reports from China and India (Prasad and Jauhari, 1963; Misra, Nauriyal, and Awasthi, 1973; Khan, Misra, and Srivastava, 1976; Yuan and Huang, 1991). This contradiction may be the result of genetic differences between cultivars, environmental conditions, or the timing of the synthetic auxin application. It is also possible that the commercial formulation used in Tipimon improved the efficacy of the auxin, compared with the formulations used in the other countries.

Several years ago, 2,4,5-T was withdrawn from the market because it contained a carcinogenic contaminant. It was suspected that other phenoxy auxins, including 2,4,5-TP, were similarly contaminated (Arteca, 1996) and would be withdrawn (Gaash, David, and Doron, 1993). However, Arteca (1996) reported that the carcinogen had been eliminated in the current methods of phenoxy auxin synthesis, so the withdrawal of 2,4,5-TP is no longer expected.

In the meantime, Stern and Gazit (1997) tested a large number of synthetic auxins in a search for an alternative to 2,4,5-TP in controlling fruitlet drop in 'Mauritius'. They found that the synthetic auxin Maxim (manufactured by Dow Elanco, Madrid), sold in tablet form and containing 10 percent 3,5,6-trichloro-2-pyridyl-oxyacetic acid (3,5,6-TPA), was effective in reducing abscission at 50 ppm (see Figure 8.3).

FIGURE 8.3. Pattern of 'Mauritius' Lychee Fruitlet Abscission in Control Inflorescences and After Spraying with 2,4,5-TP (100 ppm) or 3,5,6-TPA (50 ppm) at the Stage When Fruitlets Weighed About 2 g

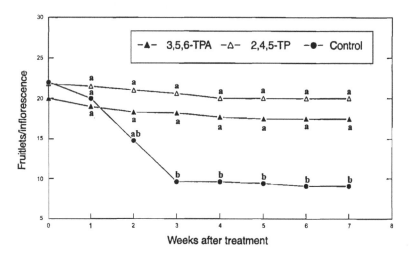

Source: Stern and Gazit, 1997, pp. 659-663.

Note: Data are the means of 16 inflorescences. Mean separation in each week by Duncan's multiple range test, $P = 0.05$.

Recently, Stern and Gazit (2000) found that the highest yield in 'Mauritius' and 'Floridian' was obtained after spraying with 2,4,5-TP (100 ppm) and 3,5,6-TPA (20 ppm) several days later (see Figure 8.4).

CONCLUSION

Two synthetic auxins (2,4,5-TP and 3,5,6-TPA), as formulated in two commercial products (Tipimon and Maxim, respectively), were found in Israel to consistently and significantly reduce lychee fruitlet abcission and increase yield in the three commercial cultivars Mauritius, Floridian, and Kaimana. Both auxins are now being used routinely in commercial lychee orchards in Israel. In adapting this treatment to other lychee growing regions in the world, it should be

FIGURE 8.4. Effect of Tipimon (100 ppm 2,4,5-TP) and Maxim (20 ppm 3,5,6-TPA) Spraying of Young 'Mauritius' and 'Floridian' Plots on Marketable Fruit Yield

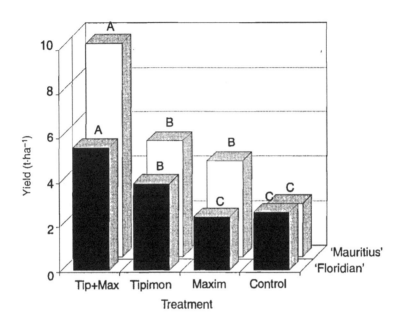

Source: Stern and Gazit, 1997, pp. 659-663.

Note: Data are the means of 200 trees per treatment. Mean separation in each cultivar by Duncan's multiple range test, $P = 0.05$.

emphasized that timing of the spraying is crucial and may be different for other cultivars. Comprehensive trials should be conducted in other countries and on different cultivars to realize the potential of these and other auxins to significantly increase lychee yield.

REFERENCES

Addicott, F.T. (1982). *Abscission.* Berkeley, CA: University of California Press.
Arteca, R.N. (1996). *Plant Growth Substances: Principles and Applications.* New York: Chapman and Hall Press.
Batten, D.J. (1986). Towards an understanding of reproductive failure in lychee. *Acta Horticulturae* 175:79-83.

Chadha, K.L. and M.S. Rajpoot (1969). Studies on floral biology, fruit set, and its retention and quality of some litchi varieties. *Indian Journal of Horticulture* 26:124-129.

Chaikittiyos, S., C.M. Menzel, and T.S. Rasmussen (1995). Floral induction in tropical fruit trees: Effect of temperature and water supply. *Journal of Horticultural Science* 69:397-415.

Chaitrakulsub, T., P. Chaidate, and H. Gemma (1988). Study of fruit development of *Litchi chinensis* Sonn. var. Hong-Huay. *Japanese Journal of Tropical Agriculture* 32:201-207.

Chandler, W.H. (1958). *Evergreen Orchards*. Philadelphia, PA: Lea and Febeger Press.

Gaash, D., I. David, and I. Doron (1993). Various auxin alternative formulations to reduce preharvest drop of apple. *Advances in Horticultural Science* 7:89-91.

Galan-Sauco, V. and U.G. Menini (1989). *Litchi Cultivation*. Rome: Food and Agriculture Organization.

Goren, R. (1993). Anatomical, physiological, and hormonal aspects of abscission in citrus. *Horticultural Reviews* 15:145-182.

Goren, R. and M. Huberman (1976). Effects of ethylene and 2,4-D on the activity of cellulase isoenzymes in abscission zone of the developing orange fruit. *Physiologia Plantarum* 37:123-130.

Groff, G.W. (1948). Additional notes upon the history of the 'Brewster' lychee. *Proceedings of the Florida State Horticultural Society* 61:285-289.

Guar, G.S. and P.N. Bajpai (1978). Some aspects of developmental physiology of the litchi fruit. *Indian Journal of Horticulture* 35:173-177.

Hayes, W.B. (1945). *Fruit Growing in India*. Allahabad, India: Kitahistan.

Hoda, M.N., N.B. Syamal, and V.S. Chhonkar (1973). Effect of growth substances and zinc on litchi fruit drop. *Indian Journal of Horticulture* 30:456-460.

Huang, H. and J. Xu (1983). The development patterns of fruit tissues and their correlative relationships in *Litchi chinensis* Sonn. *Scientia Horticulturae* 19:335-342.

Joubert, A.J. (1967). Die bloei, Embriosak-, Embrio-en Verugontwikkeling von *Litchi chinensis* Sonn. cultivar HLH Mauritius. MS Thesis. University of Witwatersrand, Johannesburg, South Africa.

Joubert, A.J. (1986). Litchi. In *Handbook of Fruit Set and Development,* S.P. Monselise (Ed.). Boca Raton, FL: CRC Press, pp. 233-246.

Kanwar, J.A., G.S. Nijjar, and M.S. Rajput (1972). Fruit growth studies in litchi (*Litchi chinensis* Sonn.) at Gurdaspur (Punjab). *Punjab Horticultural Journal* 12:146-151.

Khan, I., R.S. Misra, and R.P. Srivastava (1976). Effect of plant growth regulators on the fruit drop, size and quality of litchi cultivar Rose Scented. *Progressive Horticulture* 8:61-69.

Khan, K.S. (1929). Pollination and fruit formation in litchi (*Nephelium litchi,* Camb.). *Agricultural Journal of India* 24;183-187.

Knight, R. Jr. (1980). Origin and world importance of tropical and subtropical fruit crops. In *Tropical and Subtropical Fruits: Composition, Properties and Uses,* S. Nagy and P.E. Shaw (Eds.). Westport, CT: AVI Publishing, pp. 1-120.

Leopold, A.C. (1958). Auxin uses in the control of flowering and fruiting. *Annual Review of Plant Physiology* 9:281-310.

Leopold, A.C. and P.E. Kriedman (1975). *Plant Growth and Development*. New York: McGraw-Hill.

Liu, J. (1986). Studies on the changes of endogenous indoleacetic acid and gibberellin during litchi fruit development. MS Thesis. South China Agricultural University, Guangzhou, China.

McConchie, C.A. and D.J. Batten (1989). Floral biology and fruit set in lychee. *Proceedings of the 2nd National Lychee Seminar*, Cairns, Australia, pp. 71-74.

Menzel, C.M. (1983). The control of floral initiation in lychee: A review. *Scientia Horticulturae* 21:201-215.

Menzel, C.M. (1984). The pattern and control of reproductive development in lychee: A review. *Scientia Horticulturae* 22:333-341.

Menzel, C.M. and D.R. Simpson (1990). Performance and improvement of litchi cultivars: A review. *Fruit Varieties Journal* 44:197-215.

Misra, S.K., J.P. Nauriyal, and R.P. Awasthi (1973). Effect of growth regulators on fruit drop in litchi. *Punjab Horticultural Journal* 13:122-126.

Mustard, M.J. (1960). Megagametophytes of the lychee (*Litchi chinensis* Sonn.). *Proceedings of the American Society for Horticultural Science* 75:292-303.

Mustard, M.J., S. Liu, and R.O. Nelson (1953). Observation of floral biology and fruit setting in lychee varieties. *Proceedings of the Florida State Horticultural Society* 66:212-220.

Pivovaro, S.Z. (1974). Studies on the floral biology and the influence of growth regulators on fruit set, size and drop of *Litchi chinensis* Sonn. MS thesis. The Hebrew University of Jerusalem, Israel (in Hebrew).

Prasad, A. and O.S. Jauhari (1963). Effect of 2,4,5-trichlorophenoxyacetic acid and alpha naphthaleneacetic acid on "drop stop" and size of litchi fruits. *Madras Agricultural Journal* 50:28-29.

Samson, J.A. (1980). *Tropical Fruits*. London: Longman Press.

Shalem-Galon, M. (1980). Lychee: Fertilization, fruit set and storage. MS Thesis. The Hebrew University of Jerusalem, Israel (in Hebrew).

Shoan, S. and B.S. Dhillon (1981). Fruit drop pattern in litchi cultivars and its control by the use of auxins. *Progressive Horticulture* 13:91-93.

Singh, U.S. and R.K. Lal (1980). Influence of growth regulators on setting, retention and weight of fruits in two cultivars of litchi. *Scientia Horticulturae* 12:321-326.

Stern, R.A. (1992). Elucidation of the factors affecting litchi productivity in Israel, and the development of methods to improve its yield. PhD Thesis. The Hebrew University of Jerusalem, Israel (in Hebrew).

Stern, R.A., I. Adato, M. Goren, D. Eisenstein, and S. Gazit (1993). Effect of autumnal water stress on litchi flowering and yield in Israel. *Scientia Horticulturae* 54:295-302.

Stern, R.A. and S. Gazit (1996). Lychee pollination by the honeybee. *Journal of the American Society for Horticultural Science* 121:152-157.

Stern R.A. and S. Gazit (1997). Effect of 3,5,6-trichloro-2-pyridyl-oxyacetic acid on fruitlet abscission and yield of 'Mauritius' litchi (*Litchi chinensis* Sonn.) *Journal of Horticultural Science* 72:659-663.

Stern, R.A. and S. Gazit (1998). Pollen viability in lychee. *Journal of the American Society for Horticultural Science* 123:41-46.

Stern, R.A. and S. Gazit (1999). The synthetic auxin 3,5,6-TPA reduces fruit drop and increases yield in 'Kaimana' litchi. *Journal of Horticultural Science and Biotechnology* 74: 203-205.

Stern, R.A. and S. Gazit (2000). Effect of 2,4,5-TP and 3,5,6-TPA on litchi fruit size and yield. *Hort Science* (in press).

Stern, R.A., S. Gazit, R. El-Batsri, and C. Degani (1993). Pollen parent effect on outcrossing rate, yield, and fruit characteristics of 'Floridian' and 'Mauritius' lychee. *Journal of the American Society for Horticultural Science* 118:109-114.

Stern, R.A., J. Kigel, E. Tomer, and S. Gazit (1995). 'Mauritius' lychee fruit development and reduced abscission after treatment with the auxin 2,4,5-TP. *Journal of the American Society for Horticultural Science* 120:65-70.

Stern, R.A., D. Eisenstein, H. Voet, and S. Gazit (1996). Anatomical structure of two-day-old litchi ovules in relation to fruit set and yield. *Journal of Horticultural Science* 71:661-671.

Stern, R.A., D. Eisenstein, H. Voet, and S. Gazit (1997). Female 'Mauritius' litchi flowers are not fully mature at anthesis. *Journal of Horticultural Science* 72:19-25.

Stern, R.A., M. Meron, A. Naor, R. Wallach, B. Bravdo, and S. Gazit (1998). Effect of fall irrigation level in 'Mauritius' and 'Floridian' lychee on soil and plant water status, flowering intensity, and yield. *Journal of the American Society for Horticultural Science* 123:150-155.

Stern, R.A., M. Nadler, and S. Gazit (1997). 'Floridian' litchi yield is increased by 2,4,5-TP spray. *Journal of Horticultural Science* 72:609-615.

Tindall, H.D. (1994). Sapindaceous fruits: Botany and horticulture. *Horticultural Reviews* 16:143-196.

Veera, S. and R.C. Das (1974). Effect of 2,4-D, NAA, GA and 2,4,5-T on initial set, retention and growth of fruits in litchi, var. Muzaffarpur. *Horticultural Advances* 9:11-13.

Verma, S.K., B.P. Jain, and S.R. Dass (1981). Preliminary studies on the evaluation of the effect of growth substances with minor elements in controlling fruit drop in litchi. *Haryana Journal of Horticultural Science* 10:4-10.

Weaver, R.J. (1972). *Plant Growth Substances in Agriculture.* Davis, CA: University of California Press.

Yuan, R. and H. Huang (1988). Litchi fruit abscission: Its patterns, effect of shading and relation to endogenous abscisic acid. *Scientia Horticulturae* 36:281-292.

Yuan, R. and H. Huang (1991). Effect of NAA, NAA plus nucleotides on fruit set of lychee. *Australian Lychee Yearbook* 1:46-50.

Chapter 9

Chemical Bloom Thinning
of Pome and Stone Fruits

Renae E. Moran
Stephen M. Southwick

INTRODUCTION

Apples, pears, and most of the stone fruits have the potential for high fruit set, leading to large crops of small fruit size. The effects of overcropping include small fruit size, reduced return bloom, delayed maturity, poor fruit quality, reduced shoot growth, reduced cold hardiness, reduced pricing, and possible lower economic return. For these reasons, crop load management is essential to commercially successful tree fruit production. Pome and stone fruit trees have traditionally been thinned by hand after fruit set because it is less risky and more discriminating, although very costly. Apples (*Malus domestica* Borkh.), European and Asian pears (*Pyrus communis* L. and *Pyrus ussuriensis* Maxim), peaches and nectarines [*Prunus persica* (L.) Batsch], apricots (*Prunus armeniaca* L.), plums (*Prunus salicina* L.), and prunes (*Prunus domestica* L.) benefit from thinning (Lilleland, 1950; Fitch, Ramos, and Yeager, 1972; Williams, 1979; Sibbett and Martin, 1982; Webster and Andrews, 1985; Gu and Lombard, 1994). Thinning of cherries (*Prunus avium* L. and *Prunus cerasus* L.) is not often practiced, since overset tends to be a minor problem. This chapter summarizes past research on bloom-thinning strategies of the

The authors wish to acknowledge the critical review of Kitren G. Weis.

223

major tree fruit crops, which includes apples, pears, peaches, nectarines, apricots, plums, and prunes.

Bloom thinning can increase fruit size, yield, and return bloom more than later thinning dates, so methods have been developed that allow growers to manage crop load before or during bloom. Bloom-thinning agents can be used from before the period of floral induction through petal fall. Bloom-thinning strategies include prevention of initiation, promotion of flower bud death, prevention of fertilization, and promotion of flower abscission. Hand thinning has been, and still is, the most relied-upon strategy to reduce fruit numbers, but, due to the high cost and lack of labor, is impractical at bloom. Chemical bloom thinning became a commercial practice in the 1950s with the introduction of Elgetol, which is no longer used because it is unpredictable and can be phytotoxic. Growth regulators, caustic chemicals, and pesticides are currently being used or tested as bloom thinners.

A thinning agent must have certain traits to be acceptable. Consistent or predictable thinning from year to year is important to avoid under- or overthinning. A thinning agent should reduce crop load to a predictable amount each time it is used. At the moment, this is difficult to achieve with any thinning agent used at bloom. Consequently, knowing the origin of variation in thinning response is important for proper utility of each product. Uniformity of spacing is an added bonus so that fruits are not clustered within the canopy, since this can reduce fruit size, but a thinning agent can be successful even if fruit thinning is nonuniform. A thinning agent should be effective for at least a one- to two-week period of time so the opportunity to apply the chemical is not easily missed. A thinning agent should not cause any adverse effects on fruit quality or tree health in the current and following seasons. A thinning agent should be cost-effective, especially when combined with follow-up hand thinning, and have little or no toxicity or residual presence. When bloom thinners do not sufficiently reduce crop load, they can be followed by a consistent postbloom thinner, but, so far, postbloom thinners are available for only apples and pears.

REASONS FOR BLOOM THINNING

The advantages of bloom thinning over postbloom thinning include maximizing yield and fruit size relations, consistent return bloom, advanced fruit maturity, and greater cold hardiness. The goal of thinning is to carry the greatest number of fruit to harvest without compromising yield, fruit size, fruit quality, or return bloom, and this is done by thinning as early as possible.

Fruit thinning has been performed by hand after fruit set because of the lack of consistent chemical thinning. Currently, many bloom thinners are available for both pome and stone fruits, but many growers are reluctant to thin before fruit set because of overthinning fears. Postbloom thinning of peaches is practiced in the eastern United States due to the dangers of frost, even though bloom thinning can increase flower bud hardiness (Byers, Carbaugh, and Presley, 1990). However, an increasingly competitive market necessitates a minimum fruit quality if a grower is to profit, so the need for bloom thinners has increased.

Size and Yield

To maximize fruit size and yield, the opportune time for thinning is during or just after bloom (Farley, 1923; Dorsey and McMunn, 1927; Shoemaker, 1933; Tukey, 1936; Knight, 1978; Johnson, 1994). When thinning is done at bloom, yield can be increased because of an increase in fruit size not seen with thinning at later dates (Tukey and Einset, 1938; Havis, 1962). Delay in thinning diminishes the impact on fruit size (Weinberger, 1941; Denne, 1960; Glenn et al., 1994; Goffinet, Robinson, and Lakso, 1995). With bloom thinning, fewer fruits need to be removed from the tree to get the same size as would occur with later but increased thinning (Sharples, 1968). Fruit size is generally determined early in the season (by 50 days after full bloom [DAFB]), so small fruits will remain small up to harvest (Williams and Edgerton, 1981). Selective thinning agents increase fruit size by removing the smaller fruit, resulting in an increase in mean fruit size (Childers, Morris, and Sibbett, 1995). However, this does not explain how nonselective thinners, such as mechanical rope curtains, increase fruit size (Baugher et al., 1991). When thinning occurs at or near bloom, cell division occurs with less competition, and fruit then may have greater cell numbers (Goffinet, Robinson, and Lakso, 1995). Larger fruits typically have more cells rather than larger cells compared to smaller fruit (West-wood and Billingsley, 1967; Quinlan and Preston, 1968). Bloom thinning reduces competition in the early phases of fruit growth so as not to waste nutrients and carbohydrates used by fruit that would ulti-mately be removed if postbloom thinning were practiced. Thinning typically reduces total fruit yield, but bloom thinning minimizes this loss when overthinning does not occur (Shoemaker, 1933; Knight and Spencer, 1987; Byers and Marini, 1994; Lichou et al., 1997). Bloom thinning can result in greater peach fruit size and may allow growers to leave more fruit on the tree, thereby increasing yield com-pared to postbloom thinning (Bobb and Blake, 1938; Southwick et al., 1995).

Maturity, Firmness, and Soluble Solids

In addition to increasing fruit size, bloom thinning can hasten or delay maturity depending on the thinning agent. Bloom thinning of apple and peach advanced maturity of apple compared to thinning three to five weeks later, but this may have also been due to a lighter crop on bloom-thinned trees (Quinlan and Preston, 1968; Stembridge and Gambrell, 1971; Johnson, 1994; Lichou et al., 1997). However, when peach crop load was nearly equal for all thinning times, matu-rity was advanced by bloom thinning compared to thinning two to twelve weeks later (Havis, 1962). When gibberellic acid (GA_3) is used to reduce crop load, fruit maturity, as measured by ground color, may be delayed (Brown, Crane, and Beutel, 1968) or unaffected compared to postbloom thinning (Southwick, Yeager, and Zhou, 1995; Southwick et al., 1995; Southwick, Yeager, and Weis, 1997). But when crop load was significantly reduced in the season following application, maturity was advanced (Southwick, Yeager, and Zhou, 1995). Firmness, which is important to postharvest storage life, is affected by GA_3. Preharvest application of GA_3 can increase fruit firmness of apricot, peach, plum, prune, and cherry in the same sea-son that it is applied (Facteau, Rowe, and Chestnut, 1985; Southwick and Fritts, 1995; Southwick and Yeager, 1995; Southwick, Yeager, and Zhou, 1995; Southwick et al., 1995; Southwick, Yeager, and Weis, 1997). In some cases, firmness was not increased by preharvest application of GA_3, but this was an exception rather than the rule (Southwick and Fritts, 1995; Southwick, Yeager, and Weis, 1997). Postharvest application did not increase firmness of apricots in the following season (Southwick and Yeager, 1995; Southwick, Yeager, and Zhou, 1995). Firmness of apple was increased by bloom thinning with endothall at concentrations above 1.25 milliliters per liter (ml·liter[-1]) compared to hand thinning twenty days later (Bound and Jones, 1997). Thidiazuron, another bioregulator thinning agent, increased firmness, but the effect was inconsistent (Greene, 1995). Other fruit quality factors, such as anthocyanin development and soluble solids content, of peach are reported to be unaffected by bloom thinning (Fallahi et al., 1992; Byers and Marini, 1994; Southwick, Weis, and Yeager, 1996; Southwick, Yeager, and Weis, 1997); however, in some cases, soluble solids were increased by bloom thinning, but this may have been due to a difference in crop load rather than to date of thinning (Southwick and Fritts, 1995; Southwick, Yeager, and Zhou, 1995; Bound and Jones, 1997).

Return Bloom

Alternate bearing, in which a light crop load follows a year of heavy crop load, occurs in apples, pears, apricots, and prunes. If not thinned properly, most trees will produce less fruit in the following season, resulting in inconsistent cropping. A delay in the date of thinning reduces the amount of return bloom in apple (Harley et al., 1942) and peach (Tukey and Einset, 1938; Havis, 1962; Byers and Lyons, 1985; Fallahi et al., 1992; Southwick et al., 1998). For apple, the presence of fruit is the major reason for alternate bearing. More specifically, it is the GAs from the seeds that inhibit flower bud formation (Chan and Cain, 1967; Hoad, 1978; Williams and Edgerton, 1981). Return bloom is reduced when too much fruit remains on the tree past three to eight weeks after bloom (Chan and Cain, 1967; Luckwill, 1970). Alternate bearing in stone fruits is less well understood. Not all thinning agents will result in increased return bloom. The goal of thinning with GAs is to manage crop load by reducing return bloom (Bradley and Crane, 1960; Clanet and Salles, 1976). Thidiazuron reduced return bloom of apple instead of increasing it, even though it was used to promote fruit abscission (Greene, 1995).

Cold Hardiness

Bloom thinning with GA_3 or by hand can increase peach bud cold hardiness in the current season (Proebsting and Mills, 1964; Edgerton, 1966; Corgan and Widmoyer, 1971; Painter and Stembridge, 1972) or the season following bloom thinning (Chandler, 1907; Edgerton, 1948; Byers and Marini, 1994). The exact mechanism of how bloom thinning or GA_3 increases flower bud cold hardiness has not yet been determined, but it appears to be related to the increased number of flower buds or to bloom delay (Edgerton, 1966; Byers and Marini, 1994). Bloom thinning peach by hand resulted in buds on the lower half of shoots being more cold hardy than on trees thinned at a later date, but buds on the upper half of shoots did not differ in hardiness (Byers and Marini, 1994). Greater flower survival of a freeze in the year following bloom thinning was attributed to increased number of flower buds on shoots from bloom-thinned trees (Byers and Marini, 1994). Gibberellins, which thin by inhibiting floral initiation, can also increase flower bud hardiness in late winter (Proebsting and Mills, 1964; Stembridge and LaRue, 1969; Corgan and Widmoyer, 1971; Taylor and Geisler-Taylor, 1998). The greater cold hardiness was attributed to delayed development of flower primordia, since it was the smaller buds with partially developed primordia that survived the coldest temperatures (Edgerton, 1966). However, another report (Proebsting and Mills, 1964) indicated that the most retarded buds were not always the most hardy. Bloom thinning may benefit fruit production in regions where spring freezes occur frequently.

STRATEGIES OF CROP LOAD MANAGEMENT

The four main strategies of bloom thinning, outlined in Table 9.1, differ by the stage of reproductive development that is targeted, the chemicals used, and their mode of action. These strategies include the prevention of floral initiation, promotion of flower bud death, prevention of fertilization, and promotion of flower or fruit abscission. Most of these strategies involve the use of chemicals, and none is as consistent or selective as hand thinning, but chemical compounds are preferred because of their potential for cost reduction and speedy results. Plant bioregulators, such as GA_3, prevent floral initiation, and naphthaleneacetic acid (NAA) and ethephon increase fruit abscission (Burkholder and McCown, 1941; Gur, Harcabi, and Breuer-Mizrahi, 1993; Williams, 1993b; Southwick, Yeager, and Zhou, 1995; Yokota et al., 1995; Marini, 1996). The promotion of flower bud death is accomplished with dormant petroleum or vegetable oil applied approximately one month before bloom (Call and Seeley, 1989; Deyton, Sams, and Cummins, 1992; Myers, Deyton, and Sams, 1996). Caustic chemicals can be used to prevent flower fertilization (Byers and Lyons, 1985; Myers, King, and Savelle, 1993; Lichou, Jay, and Prost, 1994; Costa et al., 1995). Not all of these strategies are used commercially, but all have been shown to work in research trials (see Table 9.1).

TABLE 9.1. Chemical Bloom-Thinning Strategies for Pome and Stone Fruits

Strategy*	Available chemicals	Crop	Application time	Advantages	Limitations
Prevention of floral initiation	GA_3, GA_{4+7}	Stone fruits	Late May through early July of the season before cropping	Fruit firmness increased in season of application Long period of sensitivity Applied in summer when temperature is more consistent than spring Increased cold hardiness	Occurs before bloom or crop size is known
Promotion of flower bud death	Petroleum or soybean dormant oils	Peach	Late Jan. To mid-Feb., one to two months before bloom	Increased cold hardiness Long period of sensitivity Inexpensive Nontoxic to mammals	Cannot be used in freezing temperatures or very dry weather
Prevention of fertilization	Surfactants, fertilizers, Endothall, pelargonic acid	Pome and stone fruits	During and up to full bloom	Applied when bloom conditions are known	Foliar phytotoxicity Fruit blemishing Short effective time
Promotion of abscission	NAA, NAD, ethephon, carbaryl, oxamyl, BA	Pome fruits	Petal fall or postbloom	Compounds are already well-tested	Inconsistent in the bloom period

* Authors are cited in text.

Prevention of Floral Initiation

Floral initiation and its inhibition take place in the season prior to bloom, so chemicals that prevent this from occurring are applied in the season preceding cropping. GAs inhibit floral initiation of pome and stone fruits and, if applied properly, can be used for crop load management (Hull and Lewis, 1959; Bradley and Crane, 1960). The use of GA_3 on stone fruits has been developed for commercial use in California, since it has been shown to work consistently (Southwick and Fritts, 1995; Southwick, Yeager, and Zhou, 1995; Southwick, Yeager, and Weis, 1997), but research on the use of GAs for pome fruit thinning has lagged behind. Flower bud induction is not simultaneous for all flower buds, so application may possibly be timed to prevent initiation of some buds while allowing others to form (Li et al., 1989). To reduce bloom of peach, GA_3 was more effective than GA_{4+7}, but GA_{4+7} increased fruit size more than GA_3 (Gur, Harcabi, and Breuer-Mizrahi, 1993). This compound has two modes of action, depending on when it is applied. GA_3 (Ralex, Abbott Labs, North Chicago, Illinois) reduces flowering in *Prunus* species (spp.) by interfering with the induction process when applied before late summer (Bradley and Crane, 1960), but it is phytotoxic to some of the flower buds when applied in late summer or early fall, after induction has occurred (Painter and Stembridge, 1972). The exact effect of GAs on the induction process is not clearly known, but evidence for apple suggests that they indirectly inhibit flower bud formation by lengthening the plastochron (Fulford, 1966; Luckwill, 1974). It appears that a critical number of nodes needs to be initiated within the bud, typically sixteen to twenty, for induction to occur, but when the plastochron is lengthened, the critical number of nodes may not be attained, so the bud remains vegetative (Fulford, 1966; Luckwill, 1974; McLaughlin and Greene, 1991). This process is described in greater detail by Faust (1989). Luckwill and Silva (1979) reported that GA_3 did not affect the plastochron of all buds, but rather the number of buds that experienced a shortening of the plastochron approximately ten weeks after bloom. Buds that became reproductive showed a shortening of the plastochron two weeks prior to the first appearance of floral primordia, but buds that remained vegetative exhibited an increase in the plastochron (Luckwill and Silva, 1979). At the proper dose, GAs have an "all or nothing" effect on initiation, so flowers are either fully developed or not developed at all

(Bradley and Crane, 1960; Clanet and Salles, 1976; Tromp, 1982; Li et al., 1989).

Timing of application is important to prevent floral initiation. After June, thinning with GA_3 becomes increasingly difficult (South-wick, Weis, and Yeager, 1996; Southwick, Yeager, and Weis, 1997; Taylor and Geisler-Taylor, 1998). Gibberellin is more effective in reducing the number of peach flower buds from May through July than before or after this time (Li et al., 1989; Byers, Carbaugh, and Presley, 1990; Gur, Harcabi, and Breuer-Mizrahi, 1993; Ward, 1993; Southwick et al., 1995), with June being the most effective time for freestone peach in the midwestern United States (Taylor and Geisler-Taylor, 1998). Corgan and Widmoyer (1971) reported August was a more effective time for thinning peach than was July or September when applications were made to individual limbs rather than whole trees. In Israel, mid-May or six weeks after full bloom was reported to be the most effective time compared to later dates for the cultivar Rhodes (Gur, Harcabi, and Breuer-Mizrahi, 1993). It is not known if location or cultivar was the cause of this variability, but cultivar variability has been demonstrated (Brown, Crane, and Beutel, 1968; Southwick and Fritts, 1995; Taylor and Geisler-Taylor, 1998). Thinning of 'Cresthaven' was more time dependent than of 'Redhaven' in the Midwest of the United States (Taylor and Geisler-Taylor, 1998). In California, 'Halford', 'Loadel', and 'Palora' varied in the amount of thinning, but flower bud density was also dif-ferent for these cultivars (Brown, Crane, and Beutel, 1968). This also occurred in another study (Southwick and Fritts, 1995) in which 'Carson' had lower bud density but greater reduction in flowering than 'June Lady', 'Elegant Lady', 'Queen Crest', and 'Andross'. Cultivar variation in response to a given concentration of GA_3 was also attributed to a greater proportion of short shoots on 'Redwing' than on 'Redskin' or 'Suncrest', since short shoots were much less sensitive to date of application than flowering on long shoots (Li et al., 1989). With apricots in California, variation among cultivars appeared to be small. Consistent results were obtained with 'Patterson', 'Royal Blenheim', 'Modesto', 'Katy', 'Tomcot', 'Goldbar', 'Gold-strike', and 'Improved Flaming Gold' when GA_3 was applied in the mid-May to early June period in most seasons (Southwick and Fritts, 1995; Southwick, unpublished data). Application in late summer can also reduce peach flower bud number by killing buds rather than pre-venting floral initiation (Painter and Stembridge, 1972), but this type of thinning with GA_3 has not been extensively evaluated for horticultural use. Time of application affects where on the shoot inhibition occurs in the case of peaches, which bear fruit laterally on one-year-old shoots. Flowering on the basal portion of shoots was inhibited by application early in the season, and on the distil portion of the shoot, by later application (Beres, 1962; Li et al., 1989; Byers, Carbaugh, and Presley, 1990; Ward, 1993; Southwick et al., 1995). Nodes on the basal portion of shoots initiate flower buds before nodes on the terminal portion of shoots (Dorsey, 1935), and this may be the reason why later applications have little effect on the basal portion of the shoots. Timing of sprays for selective, positional thinning may be advantageous for better fruit size, since buds borne near the terminal portion of the shoots produce the largest fruit (Spencer and Couvillon, 1975; Byers, Carbaugh, and Presley, 1990).

The effective time for application of GA_3 to apricots, prunes, and plums is similar to peaches. Apricots and plums were successfully thinned by application in late May to July (Southwick, Yeager, and Zhou, 1995; Southwick, Yeager, and Weis, 1997). In California, the effective period for flower reduction in 'Patterson' apricot is from May 20 through the second week of July (Southwick and Yeager, 1991, 1995). Prunes were thinned when GA_3 was applied three or seven weeks after bloom (Beres, 1962), and plums, four to six weeks after bloom (Hartmann, 1984). As with peach, low rates inhibited floral initiation of apricot when applied before it occurred, but after initiation had occurred, higher rates thinned by killing flower buds (Weis, unpublished data). During three years of treating the same 'Patterson' apricot trees, consistent thinning occurred, with no phytotoxicity and no adverse carryover effects (Southwick, Yeager, and Weis, 1997).

In addition to being time dependent, the use of GA_3 for flower reduction is concentration dependent, but the optimum concentration may vary with date of application. Application in May was effective only at a rate of 300 milligrams per liter (mg·liter^{-1}), but in June, 25 mg·liter^{-1} was sufficient to cause 27 percent reduction in the number of peach flower buds (Clanet and Salles, 1976). Concentrations of 50, 75, 100, and 120 mg·liter^{-1}, when applied to peach in June, caused a nearly equal reduction in flowering, but in early July, these concentrations caused a linear reduction, and in late July, they had no effect, indicating decreasing sensitivity after June (Southwick et al., 1995). Variability exists in the response to concentration, and it is not yet known whether this is due to variation in location, cultivar, or flower bud density. In the United States, 50 mg·liter^{-1} at the proper time is sufficient to reduce crop load and eliminate the need for hand thinning (Southwick et al., 1995; Taylor and Geisler-Taylor, 1998), but in Israel, 100 to 200 mg·liter^{-1} reduced, but did not eliminate, hand- thinning time (Gur, Harcabi, and Breuer-Mizrahi, 1993). The optimum concentration for thinning is also affected by bud loss due to inadequate winter chilling, so 50 to 75 mg·liter^{-1} has been recommended; 100 mg·liter^{-1} is considered too high because of the potential for bud loss in 'Patterson' apricot in California (Southwick, Yeager, and Weis, 1997). Similar concentrations caused thinning of plums, cherries, and apricots, but response to a given concentration varied between species and cultivar (Beres, 1962; Brown, Crane, and Beutel, 1968; Facteau, Rowe, and Chestnut, 1989; Southwick and Fritts, 1995; Southwick, Yeager, and Zhou, 1995; Southwick, Yeager, and Weis, 1997). The addition of a surfactant reduced the needed concentration (Gur, Harcabi, and Breuer-Mizrahi, 1993), but consistent results were attained without the use of a surfactant (Southwick, Yeager, and Weis, 1997).

The advantages of using GA$_3$ are a period of three to four weeks during which it is effective, a consistent and even reduction of crop load, and low toxicity. An objective of bloom thinning is to reduce or eliminate the need for hand thinning, and several studies have shown that GA$_3$ does this (Edgerton, 1966; Gur, Harcabi, and Breuer-Mizrahi, 1993; Southwick, Yeager, and Zhou, 1995; Southwick et al., 1995; Southwick, Yeager, and Weis, 1997). Another objective is to increase fruit size. Because GA$_3$ is a bloom thinner, it is expected to increase fruit size compared to hand thinning, and this has occurred in several instances (Edgerton, 1966; Byers, Carbaugh, and Presley, 1990; Gur, Harcabi, and Breuer-Mizrahi, 1993; Southwick, Yeager, and Zhou, 1995; Southwick, Yeager, and Weis, 1997).

The use of GAs for thinning pome fruits has not been developed for commercial use because of the availability of other thinning agents and the reluctance to thin before fruit set occurs. As with stone fruits, the type of GA can affect the amount of thinning. In contrast to stone fruits, GA$_{4+7}$ was more effective than GA$_3$ in reducing bloom of apple (Tromp, 1982). This difference was slight in other studies (Greene, 1981; Wertheim, 1982). GA$_7$ alone reduced bloom as much as GA$_{4+7}$, indicating that it is the GA$_7$ that caused reduction in return bloom (Tromp, 1982). Few extensive studies have been conducted on timing, but it appears that GAs must be applied from bloom to four weeks after bloom to affect flowering of spurs, and in July to reduce bloom on new wood of apple (Luckwill and Silva, 1979; Tromp, 1982). The lack of flower reduction on new wood by application soon after bloom was most likely due to the fact that extension shoots had not fully formed at that time (Luckwill and Silva, 1979; Tromp, 1982).

Thinning of apple fruits with GAs is concentration dependent. GA_{4+7} or GA_3 at 500 mg·liter^{-1} and 300 ml·liter^{-1} GA_{4+7} reduced bloom, but 150 mg·liter^{-1} did not (Luckwill and Silva, 1979; Greene, 1981; Tromp, 1982; Li et al., 1995). When concentrations of 100 or 200 mg·liter^{-1} GA_{4+7}, with or without benzylamino purine (BA), were applied to 'Delicious' at full bloom, conflicting results on return bloom occurred (Stembridge and Morrell, 1972; Unrath, 1974). The cause of this inconsistency is unknown. Split applications did not increase the consistency of response. When 25 ml·liter^{-1} GA_{4+7} was applied five times, from 11 to 61 DAFB, or when 50 ml·liter^{-1} was applied three times, from 22 to 65 DAFB, return bloom was severely reduced (McLaughlin and Greene, 1984); however, when 40 ml·liter^{-1} was applied in four split applications beginning at the end of bloom, a small reduction in bloom occurred in two trials, but none in another two trials (Wertheim, 1982). A reduction in bloom does not always occur, and this inconsistency cannot be attributed to differences in cultivar, concentration, or timing (Stembridge and Morrel, 1972; Wertheim, 1982). Hall and colleagues (1997) reported that factors such as leaf hairiness, leaf type, and age can affect retention of applied chemicals. Factors such as these may cause inconsistencies in thinning response. Other factors that may need further investigation are the effects of temperature, spray volume, surfactants, and the amount of foliage present at the time of application.

Fewer data are available for European and Asian pears, since GA_3 studies on pear were not conducted to determine the thinning ability, but instead were aimed at determining the effect on fruit set, size, and shape. Nevertheless, it has been suggested that GA_3 be used to regulate cropping of alternate-bearing cultivars because of its ability to inhibit flower formation (Turner, 1973). When applied at the phenological stages of bud swell, pink bud, full bloom, and petal fall, the number of fruit buds was reduced by 200 or 500 mg·liter^{-1} GA_3, but not by lower concentrations (Griggs and Iwakiri, 1961). When 100 mg·liter^{-1} was applied at full bloom or petal fall to 'Bartlett' in California, return bloom was unaffected (Southwick, unpublished data), but in New York, 20 to 100 mg·liter^{-1} reduced flowering of the same cultivar (Dennis, Edgerton, and Parker, 1970). Complete inhibition of flowering of 'Conference' pear was attained with full bloom or petal fall sprays of 50 mg·liter^{-1} (Turner, 1973). It is not yet known what caused the inconsistent response to concentration, but it may be related to crop load (Dennis, Edgerton, and Parker, 1970). When GA_3 was applied in a year with a light crop load, it did not reduce return bloom (Dennis, Edgerton, and Parker, 1970). The effect of GA_3 on fruit quality may limit the use of GA_3 on pears because it may result in malformed fruit with protuberant calices (Griggs and Iwakiri, 1961). Little is known about the GA thinning effects on fruit size and yield of apple and pear, but research on this topic is necessary if GAs are to be used as thinning agents.

Daminozide and maleic hydrazide have been tested to determine their effect in the prevention of floral initiation in peach, but flower bud formation was slightly increased instead of decreased (Edgerton, 1966). Thidiazuron may have the potential to manage crop load by preventing floral initiation (Greene, 1995), but it has not been tested in this capacity.

Promotion of Flower Bud Death

Although dormant soybean oil is not a growth regulator, it can kill flower buds before bloom and may have potential as a bloom thinner (Myers, Deyton, and Sams, 1996). If adopted as a thinning agent, it would give growers an additional opportunity to thin stone fruits prior to bloom. Dormant soybean oil did not thin apple flower buds at concentrations of 15 percent or less (Deyton, unpublished data). Dormant petroleum oil is often used for control of scale, aphid, and mite eggs. The dose is generally 2 percent to 6 percent v/v, but growers in California also use it to facilitate rest breaking of stone fruits. When applied in late January or early February, peach flower bud death resulted that was concentration dependent from 0 percent to 10 percent soybean or petroleum oil (Call and Seeley, 1989; Deyton, Sams, and Cummins, 1992; Myers, Deyton, and Sams, 1996). When applied at concentrations above 10 percent, shoot dieback may occur. Dormant oils have the advantage of being a prebloom thinner and may result in greater fruit size than later thinners, but this remains to be demonstrated. Oils have minimal mammalian toxicity, and registration for use may be easier than for other products. It is recommended that dormant oil not be applied when temperatures are expected to be below freezing or in very dry conditions. Its use is still in the experimental stage and has only been tested on peaches, cherries, and apples. Bloom delay, which postpones the loss of cold hardiness, can occur with application of dormant oil following the end of endodormancy (Call and Seeley, 1989; Myers, Deyton, and Sams, 1996). This can be an added benefit in areas where spring freezes occur.

Prevention of Pollination or Fertilization

Pollination and fertilization are required for adequate fruit set for pome and stone fruit crops. Crop load can be managed by preventing fertilization of some flowers while allowing others to be fertilized. This random fertilization depends on the compound striking flowers in the sensitive stage. Dinitro compounds, surfactants, fertilizers, herbicides, and other compounds have been tested as caustic bloom thinners. The mode of action of caustic sprays is dessication of the stigma, peduncle phytotoxicity, reduced pollen viability, and decreased fertilization (MacDaniels and Hildebrand, 1939; Byers and Lyons, 1985; Costa et al., 1995; Williams et al., 1995). Bloom desiccants work best during the bloom period, which can be a very short time, and this may limit the opportunity to apply the chemical. A protracted bloom can make timing of single applications difficult. These compounds may result in inconsistent thinning, which stems from changes in sensitivity of the reproductive organs as they pass through bloom, the occurrence of extended bloom periods (Zilkah, Klein, and David, 1988; Bound and Jones, 1997), and sprayer technology and volume (Byers and Lyons, 1982). However, the effectiveness of desiccating chemicals is less likely to be temperature dependent than the postbloom bioregulators currently available (Brown, Crane, and Beutel, 1968) so thinning consistency may be less weather dependent but more time dependent than with the postbloom thinners.

Elgetol (4,6-dinitro-ortho-cresylate) was used as a bloom thinner before other compounds became available (Williams, 1979). The cost of reregistration and its inconsistency when applied before rain has prevented its continued use after 1989 in many areas of the world.

Sulfcarbamide (Wilthin, D-88, or moncarbamide dihydrogensulfate, Unocal Corporation, West Sacramento, California), a surfactant, has been tested on apple and peach, but little or no research has involved other fruit crops. Sulfcarbamide, at a rate of 0.25 percent to 0.50 percent, is recommended for apple thinning no later than 80 percent bloom to get a 25 to 50 percent reduction in fruit set (Williams, 1993a). After 80 percent open bloom, there is a risk of russeting apple, especially if it follows a mineral element spray (Williams, 1993a). Sufficient crop load reduction of peach and nectarine was attained with application at full bloom (Klein and Cohen, 1995) or at 95 percent open bloom (Myers, King, and Savelle, 1993). A similar concentration as recommended for apple, 0.25 percent to 0.50 percent, worked well for peaches in the southern United States (Myers, King, and Savelle, 1993), but in Israel, a higher concentration, 3 percent, is needed to get adequate thinning of nectarine (Klein and Cohen, 1995). Inconsistent results have been obtained with 'Loadel' cling peach and 'Improved French Prune' in California (Southwick, unpublished data). Slight foliar phytotoxicity, but no fruit blemishing, occurred in some tests on peaches (Myers, King, and Savelle, 1993; Klein and Cohen, 1995). The occurrence of leaf injury was dose dependent and increased with a slow rate of drying (Curry and Williams, 1991). Apple cultivars vary in their response to this compound. A rate of 0.5 percent underthinned 'Fuji', but overthinned 'Delicious' and 'Gala' (Williams, 1993a). As with apple, peach cultivar variability occurred when three cultivars were tested in Israel (Klein and Cohen, 1995). In years with insufficient chill, sulfcarbamide did not cause excessive thinning (Klein and Cohen, 1995). However, a spring freeze did cause overthinning and greater loss of peach fruit compared to untreated trees (Myers, King, and Savelle, 1993), and this could limit its use in areas where spring freezes occur frequently.

Another surfactant that has been tested for thinning of peach is Armothin (hydroxypolyoxyethylene/polyoxypropylene ethyl alkylamine; Akzo-Nobel Chemicals, Incorporated, Chicago, Illinois). Similar to sulfcarbamide, Armothin worked best when applied at 40 to 70 percent open bloom in France (Lichou, Jay, and Prost, 1994, 1995; Lichou et al., 1997), and from 40 to 80 percent full bloom for thinning 'Loadel' cling peach in California (Southwick, Weis, and Yeager, 1996; Southwick et al., 1998). The concentration that causes adequate thinning is 2 to 3 percent when applied in a carrier volume of 935 or 1,000 liters per hectare (liter·ha^{-1}) (Lichou, Jay, and Prost, 1994, 1995; Southwick, Weis, and Yeager, 1996). Increasing carrier volume increased thinning response (Southwick et al., 1998). Some variability among cultivars was reported, but cultivars were not tested side by side and were sprayed at different bloom stages (Lichou et al., 1997). Cultivars that differ by harvest date are currently being tested in California (Southwick, unpublished data). Armothin at 3 percent caused phytotoxicity of freestone peaches in France (Lichou, Jay, and Prost, 1994; 1995), especially when warm temperatures followed application, but a concentration of 5 percent was needed to cause foliar phytotoxicity on clingstone peach in California (Southwick, Weis, and Yeager, 1996). Armothin did not cause fruit blemishing of peach when applied at 80 percent of full bloom, but increased the number of undersized fruit (Southwick, Weis, and Yeager, 1996). However, recent observations suggest that application of Armothin at 40 percent bloom may result in less undersized fruit at harvest when compared to sprays applied at 80 percent full bloom in peach (Southwick, unpublished data). Armothin resulted in larger fruit size at harvest compared to hand thinning (Southwick et al., 1998). Increased early season fruit size was found in another study (Southwick, Weis, and Yeager, 1996), but this difference did not occur at harvest. Greater yield was obtained in one study (Lichou et al., 1997), but not in another (Southwick et al., 1998). Armothin significantly reduced the time required for hand thinning from a mean of 30 minutes (min) per tree to 7 min (Southwick et al., 1998). Published research on other fruit crops is lacking.

Several other surfactants, X77, X45, Dupont WK, SN-50, and CC-42, have shown thinning ability on peach, but the concentration range tested was limited (Byers and Lyons, 1982; 1985). X77 and X45 did not reduce crop load enough to eliminate the need for hand thinning or increase fruit size more than hand thinning at a later date (Byers and Lyons, 1982), but Dupont WK and SN-50 increased fruit size more than hand thinning (Byers and Lyons, 1985). SN-50 thinned only one of seven peach cultivars at 15 ml·liter^{-1}, possibly due to too low a concentration, so conclusions about cultivar response cannot be made from this study (Byers and Lyons, 1984). All of these surfactants caused foliar phytotoxicity, and Dupont WK, SN-50, and CC-42 resulted in some undersized fruit (Byers and Lyons, 1982, 1985), indicating that they have similar problems as sulfcarbamide and Armothin.

The fertilizers ammonium thiosulfate (ATS), ammonium nitrate, and urea have been tested for bloom-thinning ability on peach and, to a lesser extent, on apple. ATS, which is 43 percent sulfur and 19 percent nitrogen, is best applied at 80 percent to 90 percent open bloom (Byers, 1987; Olien et al., 1995). Response was dependent on how ATS was applied. Airblast application of ATS at 30 ml·liter^{-1} in 1,170 liters·ha^{-1} carrier volume thinned fewer fruit than handgun application at the same rate and volume (Byers and Lyons, 1985). Also, greater deposit occurred when blocks of trees were sprayed than when spraying a single row of trees, so a lower concentration should be used when applying to whole blocks of trees (Byers, 1987). Application methods used for research trials may not represent full-scale commercial application; thus, recommendations should also consider how the compound is to be applied. Application with benomyl increased thinning response of peach, and application with other fungicides reduced it (Olien et al., 1995). The addition of a surfactant (Spray Aide) or changing the spray volume did not improve thinning (Byers and Lyons, 1985; Byers, 1987). Bloom thinning with ATS has been shown to increase peach fruit size, hasten maturity, and increase return bloom compared to hand thinning at a later date (Byers and Lyons, 1985; Byers, Carbaugh, and Presley, 1990). ATS appears to reduce losses due to freezing in the current season, but this needs further verification (Olien et al., 1995). An additional benefit may be less cultivar variability than other bloom desiccants (Byers and Lyons, 1984). Leaf and shoot burning and increased number of undersized fruit are problems also of ATS (Byers and Lyons, 1985; Olien et al., 1995). Multiple applications of ATS during bloom of 'Fuji' apple has shown some success under California conditions (Southwick, unpublished data). Bloom thinning with ATS has been more extensively researched than thinning with ammonium nitrate or urea. Ammonium nitrate did not thin peach as well as ATS and thinned fewer cultivars (Byers and Lyons, 1984), but when tested at a higher rate, thinning was comparable to ATS (Byers and Lyons, 1985). Urea has been shown to thin peaches when applied at or near the bloom stage (Zilkah, Klein, and David, 1988; Di Marco et al., 1992). Thinning with urea was concentration dependent, with 4 percent as the lowest effective rate in one study (Zilkah, Klein, and David, 1988), but 12 percent needed in another trial to have a measurable effect (Di Marco et al., 1992). Two nectarine cultivars were thinned optimally with 16 percent but were overthinned with 20 percent (Di Marco et al., 1992). Another study showed overthinning when 16 percent was applied (Zilkah, Klein, and David, 1988), indicating that response to this compound may be variable. Concentrations of 12 percent or greater were phytotoxic, causing shoot death (Zilkah, Klein, and David, 1988; Di Marco et al., 1992). Urea thinning did not result in even fruit distribution along shoots and required supplemental hand thinning (Zilkah, Klein, and David, 1988; Di Marco et al., 1992). Fertilizers show promise as thinning agents, are comparatively inexpensive, and are readily available.

Endothall [7-oxabicyclo-(2,2,1) heptane-2,3-dicarboxylic acid], an aquatic herbicide, also called endothallic acid (Elf Atochem, N.A., Philadelphia, Pennsylvania), has been tested on apple and peach. At 80 percent full bloom, thinning of apple was concentration dependent, from 0 to 3.0 ml·liter^{-1} for 'Delicious', with a rate of 2.0 to 2.5 ml·liter^{-1}, achieving a crop density of four to six fruit per cm^2 trunk cross-sectional area (CSA) (Bound and Jones, 1997). Trials on peaches indicate it is less consistent than with apples, since no thinning occurred in some of the years it was tested (Fallahi, 1997; Southwick, unpublished data). However, endothall at 1.0 ml·liter^{-1} thinned all seven peach cultivars tested and in some cases over-thinned (Byers and Lyons, 1984). Fruit size and crop load of apple were comparable to hand thinning 20 DAFB (Bound and Jones, 1997). Fruit size of peach was similar to hand thinning when under-thinning occurred (Byers and Lyons, 1985), but with greater thinning, fruit size was greater than hand-thinned trees in six of seven cultivars tested (Byers and Lyons, 1984). Endothall increased fruit firmness and soluble solids, but also resulted in flatter fruit shape of apple (Bound and Jones, 1997). Concentrations that thin to commercial fruit density levels caused phytotoxicity to foliage of peach and apple, but did not blemish fruit (Byers and Lyons, 1984, 1985; Bound and Jones, 1997; Fallahi, 1997).

Apples and peaches were thinned by full bloom applications of hydrogen cyanamide (Dormex, D.K. International, Incorporated, Marietta, Georgia) in two years of testing (Fallahi, 1997), but no other published research discusses this compound as a thinner. Pelargonic acid (Thinex; Mycogen, Corporation, San Diego, California) thinned apples and peaches at bloom in one of two years of testing, demonstrating an inconsistency with this chemical (Fallahi, 1997). It also caused russeting of apples (Fallahi, 1997).

A short effective time, foliar damage, russeting of apple, and cultivar variability are current problems of bloom desiccants. Most bloom desiccants, however, do not appear to be reactivated by rain, unlike Elgetol. Predictability, which is important in crop load management, may be better with some desiccants than with others.

Flower and Fruit Abscission

Most chemicals that promote abscission of flowers or fruitlets were developed for postbloom thinning, but can also thin when applied in the bloom or petal fall stages. Most of the compounds that promote abscission are growth regulators, but carbaryl and oxamyl are insecticides. The bioregulators include naphthaleneacetic acid (NAA), naphthaleneacetamide (NAD), ethephon, benzyladenine (BA), thidiazuron, and paclobutrazol. Postbloom thinners are not consistently effective on stone fruits (Murneek and Hibbard, 1947; Hibbard and Murneek, 1950). The mode of action of the abscission promoters is reported to be seed abortion (Marsh, Southwick, and Weeks, 1960), or interference of vascular transport of the flower or fruit (Harrold, 1935; Williams and Batjer, 1964; Crowe, 1965). The use of NAA, NAD, ethephon, and carbaryl on apple has been previously reviewed by Martin (1973), Williams (1979), and Miller (1988), so it will be only briefly discussed here.

NAA, an auxin that is normally used in the postbloom period, thinned effectively when applied at petal fall (Burkholder and McCown, 1941; Yokota et al., 1995; Marini, 1996). It may not consistently thin to a specific crop load, underthinning in some cases and overthinning in others (Williams, 1993b). The effective concentration varies with location, weather, cultivar, and use of pesticides or other thinning agents. At full bloom or petal fall, NAA is a potent thinner at 5 to 30 ml·liter^{-1} (Jones et al., 1989; Marini, 1996; Looney, Beulah, and Yokota, 1998), but these concentrations were less effective in Israel (Ben-Arie, 1992). Thinning with NAA is dependent on temperature, which means it must be applied under proper conditions. In addition, stresses that cause early defoliation in the previous season may increase sensitivity to NAA (Rosenberger, Meyer, and Engle, 1994). Other limitations include a greater incidence of undersized (pigmy) fruit, small fruit size compared to hand thinning, and foliar damage (Williams, 1993b; Black, Buckovac, and Hull, 1995; Yokota et al., 1995). The reduction in fruit size could be from uneven thinning leading to intraspur competition, since pigmy fruit occurred only when there was more than one fruit per spur (Black, Buckovac, and Hull, 1995). NAA does not selectively thin to one fruit per spur, which is an undesirable trait in a chemical thinner (Southwick and Weeks, 1949; Williams, 1993b). Thinning within a cluster resulted in greater fruit size than thinning entire clusters (Knight, 1980), so chemicals that result in one fruit per cluster are better than those which thin all the fruit on some clusters and leave more than one on others. Another reason for small fruit size could be the immediate reduction in fruit growth, fruit cell division, and foliar CO_2 assimilation following application (Southwick et al., 1962; Black, Buckovac, and Hull, 1995; Stopar, Black, and Buckovac, 1997). The effect on fruit size persisted up to harvest (Southwick et al., 1962). NAA is effective on pears 15 to 21 DAFB, but little published information is available on thinning during bloom or petal fall.

NAD (naphthaleneacetic acid amide; Amid-Thin, AMVAC Chemical Corporation, Los Angeles, California), a compound that causes less foliar damage, but more pigmy fruit, in some cultivars, may be useful for cultivars that are easily overthinned by NAA (Greene, 1981).

Other auxin-type compounds tested as bloom thinners of apple include dichlorprop, MCPB-ethyl, NSK-905, ethychlozate, and 4-CPA, but of these, dichlorprop, MCPB-ethyl, and NSK-905 reduced fruit set, but ethychlozate and 4-CPA had no effect (Yokota et al., 1995). MCPB-ethyl thinned as well at 80 percent full bloom as it did at petal fall (Looney, Beulah, and Yokota, 1998), indicating that response may be stable over a period of time, which is a desirable trait for a thinning agent. However, similar to NAA, MCPB-ethyl did not selectively thin to one fruit per cluster, necessitating an additional thinning agent or hand thinning (Looney, Beulah, and Yokota, 1998).

Ethephon [(2-chloroethyl) phosphonic acid], an ethylene-releasing compound, has been tested as a bloom thinner of apples, pears, prunes, and peaches. Ethephon reduced fruit set of apple when applied at full bloom (Knight, 1980; Jones et al., 1990, 1993) or at petal fall (Marini, 1996). Full-bloom application decreased fruit set more than application 14 DAFB in one study (Jones et al., 1990), but had an equal effect at both dates in another (Jones et al., 1989). When applied to peach or pear at bloom or petal fall, ethephon reduced fruit set, but application after petal fall was more effective (Stembridge and Gambrell, 1971; Knight and Browning, 1986). Ethephon was also effective on prunes (Martin et al., 1975). The response of ethephon is dependent on temperature and solution pH, so lack of consistent response between different studies is not surprising. Thinning of apple was concentration dependent, but factors such as tree age, rootstock, and cultivar can affect response to a given concentration. In three-year-old 'Red Fuji', 400 mg·liter^{-1} caused nearly complete fruit drop (Jones et al., 1989), but in five-year-old trees, 400 or 800 mg·liter^{-1} reduced fruit set to a commercial level, suggesting an effect of tree age on thinning response (Jones et al., 1990). 'Delicious' was thinned to a crop density of 3.6 fruit per cm^2 limb CSA by 400 mg·liter^{-1}, but this was the only concentration tested in the bloom period in this study (Marini, 1996). A concentration of 200 mg·liter^{-1} sufficiently reduced fruit set of 'Jonagold', 800 mg·liter^{-1} overthinned, and 1,600 mg·liter^{-1} completely eliminated fruit set (Jones et al., 1993). Thinning of 'Jonagold' was easier than thinning of 'Gala', indicating variability in apple cultivar response (Jones et al., 1993). The effect on fruit size has been inconsistent, but this may be due to whether thinning treatments were compared to unthinned or to handthinned trees, and to when hand thinning was performed. Fruit size of apple and pear was greater compared to fruit from unthinned trees (Jones et al., 1990, 1993; Martin et al., 1975). Ethephon did not increase peach fruit size compared to hand thinning in one study (Stembridge and Gambrell, 1971), but did in another in which hand thinning was late (Edgerton and Greenhalgh, 1969). Ethephon caused foliar phytotoxicity (Edgerton and Greenhalgh, 1969; Marini, 1996), reduced shoot growth of apple (Jones et al., 1993), and resulted in earlier foliar fall coloration of prune (Sibbett and Martin, 1982).

Paclobutrazol, a growth retardant and inhibitor of gibberellin biosynthesis, promoted fruitlet abscission when applied at bloom to plum (Webster and Andrews, 1985) or peach (Blanco, 1987), but did not affect final fruit set of pear (Knight and Browning, 1986). Fruit set of plum was reduced by bloom application in one of two trials, but thinning was more consistent with application after bloom (Webster and Andrews, 1985). Initial fruit set of pear was reduced by application during bloom, but had no effect on final fruit set (Knight and Browning, 1986). As with most bloom thinners, paclobutrazol increased peach fruit size more than hand thinning at a later date (Blanco, 1987). Greater fruit size also occurred with plum when compared to an unthinned control (Webster and Andrews, 1985). Paclobutrazol resulted in a "weeping" habit of peach shoots (Coston, 1986) and fruit gumming of plum (Webster and Andrews, 1985), which are limitations of this thinning agent.

Benzyladenine (6-benzylamino purine), a cytokinin, thinned apple when applied postbloom (Greene and Autio, 1989; Elving and Cline, 1993). Cytokinins have not been evaluated as thinning agents of stone fruits. The combination of BA and GA_{4+7} (Promalin, Abbott Labs, North Chicago, Illinois) did not thin apple when applied at bloom or petal fall (Stembridge and Morrell, 1972), or did so inconsistently (Unrath, 1974). BA is effective in thinning apple when applied alone (Greene and Autio, 1989) and did not thin as well when applied in combination with GA_{4+7} (Elving and Cline, 1993), so it appears the activity of BA rather than GA_{4+7} results in thinning. When compared to carbaryl, ethephon, Ethion (an insecticide), NAA, or oxymyl, BA was the least effective thinner for apple and caused thinning only when fruit diameter was 8 millimeters (mm) (Marini, 1996), so its use as a bloom thinner may not have been demonstrated. However, it has advantages over other postbloom thinners. It increases fruit size more than carbaryl or NAA (Greene et al., 1992; Elving and Cline, 1993). The increase in fruit size is attributed to a greater rate of cell division compared to fruit from trees thinned with carbaryl or NAA, even though crop load was similar (Wismer, Proctor, and Elving, 1995). Unlike NAA, BA does not reduce CO_2 assimilation rate of leaves, and this may account for the different effects of these compounds on fruit size (Stopar, Black, and Buckovac, 1997).

Thidiazuron, a bioregulator with cytokinin-like activity, reduced fruit set of 'McIntosh' apple, but not 'Delicious' or 'Empire', when applied at bloom (Greene, 1995). In addition, return bloom was reduced in 'McIntosh' and 'Delicious', but had no effect on return bloom of 'Empire' (Greene, 1995). Thidiazuron caused misshapen fruit and reduced red pigmentation, but increased fruit fresh weight, flesh firmness, and soluble solids of 'McIntosh' and the length: diamater ratio and flesh firmness of 'Delicious' (Greene, 1995). It is not known if thidiazuron is a good thinning agent since the combined effects of reduced return bloom and fruit set have not been evaluated for crop load management.

Carbaryl (1-naphthyl-N-methylcarbamate; Sevin 50W and Sevin XLR-Plus, Rhone-Poulenc) is an insecticide that has been used as a postbloom thinner of apple for years. It is not effective on stone fruits. Its use as a postbloom thinner of apple has been previously reviewed (Williams, 1979), so only its bloom-thinning capability will be discussed. Carbaryl reduced fruit set when applied from full bloom to thirty-four days after petal fall, but thinning was greater with petal fall application than before or after petal fall (Way, 1967; Williams, 1993b). Thinning response in another study (Marini, 1996) was greater at a fruit diameter of 8 mm rather than 4 mm (petal fall). Little difference in response occurred with different application dates, from 80 percent petal fall to mid-June, but fruit size was greater with earlier application for 'Bramley', 'Discovery', and 'Spartan', but 'Cox' was more effectively thinned at 12 mm diameter than at other fruit sizes (Knight and Spencer, 1987). The advantages of carbaryl are less temperature dependence than NAA (Williams, 1993b), selective thinning of the lateral fruit within a cluster, leaving the terminal fruit (Way, 1967; Knight, 1986), no foliar damage (Williams, 1993b), and low chance of overthinning in the concentration range used (Forshey, 1987). Carbaryl can kill predators of mites and may lead to high mite populations or an increased reliance on miticides (Byers, Lyons, and Horsburgh, 1982). However, a new formulation, Sevin XLR-Plus, may reduce this problem.

Oxamyl (Vydate 2L), another pesticide, thinned when applied at bloom or petal fall with similar effectiveness as carbaryl (Byers, Lyons, and Horsburgh, 1982; Marini, 1996, 1997), but caused greater russeting of 'Golden Delicious' than carbaryl (Myers, 1982). Oxamyl reduced fruit set or crop density more at a fruit diameter of 10 mm than at 4 mm, but effects on fruit size were similar between the two application dates (Marini, 1996, 1997). Oxamyl is used commercially in the eastern United States.

UNRESOLVED PROBLEMS

Commercial chemical thinning compounds should be effective on a wide range of pome and stone fruit cultivars. They should be cost-effective, consistent in response when used properly, and have a period of use that is long enough so that applications can be made at efficacious times. They should be safe to apply and safe for the environment, with little or no adverse effects on tree health or fruit quality. Current chemical thinners with uncertain outcomes are still used, but many more options are available today than several decades ago. The inhibition of floral initiation by GAs works on most or all of the stone fruit crops but is undeveloped for use on pome fruits. There are advantages to thinning with GAs, but the major limitation is that application is made when bloom or weather conditions during bloom are unknown, and this makes growers, uncomfortable. The promotion of flower bud death with dormant oil also appears to have some promise, but requires further testing before it can be used commercially. Flower desiccants are effective for a short time during the bloom period and may occasionally cause fruit and leaf damage, but the surfactants, Armothin and Wilthin, appear promising. Compounds that promote abscission are limited by their inconsistent thinning when applied in the bloom period, making them problematic for the time being.

Unresolved problems of using chemicals as bloom-thinning agents stem from a lack of a complete understanding about what causes the variation in response. Factors including application timing, weather, tree vigor, genotypic variability, pollination, and insufficient chilling are likely contributors to variation in response. To avoid overthinning, separate applications of more than one compound at reduced concentrations may be used. Underthinning at bloom is recommended due to environmental stresses of spring freezes, insufficient chill, or cool rainy conditions at bloom. Bloom thinning can be followed with another postbloom chemical or with hand thinning if needed. In most cases, several methods may have to be used to overcome the limitations of individual thinners. However, even with sufficient research to solve some of these problems, an additional obstacle is the lack of agrochemical industry interest in registering compounds and making them available to growers because of cost, limited market potential, and liability issues. These obstacles limit investments being made to develop thinning chemicals for growers. However, more chemical thinning compounds are available today than just a few years ago.

REFERENCES

Baugher, T.A., K.C. Elliot, D.W. Leach, B.D. Horton, and S.S. Miller (1991). Improved methods of mechanically thinning peaches at full bloom. *Journal of the American Society for Horticultural Science* 116:766-769.

Ben-Arie, Z. (Ed.) (1992). Spray recommendations for deciduous orchards (in Hebrew). *Israel Ministry of Agriculture Pamphlet.*

Beres, V. (1962). The effect of gibberellin on flower bud differentiation of sugar prune. MS Thesis. University of California, Davis, California.

Black, B.L., M.J. Buckovac, and J. Hull (1995). Effect of spray volume and time of NAA application on fruit size and cropping of 'Redchief Delicious' apple. *Scientia Horticulturae* 64:253-264.

Blanco, A. (1987). Fruit thinning of peach trees [*Prunus persica* (L.) Batsch]: The effect of paclobutrazol on fruit drop and shoot growth. *Journal of Horticultural Science* 62:147-155.

Bobb, A.C. and M.A. Blake (1938). Annual bearing in the 'Wealthy' apple was induced by blossom thinning. *Proceedings of the American Society for Horticultural Science* 36:321-327.

Bound, S.A. and K.M. Jones (1997). Investigating the efficacy of endothall as a chemical thinner of 'Red Delicious' apple. *Journal of Horticultural Science* 72:171-177.

Bradley, M.V. and J.C. Crane (1960). Gibberellin-induced inhibition of bud development in some species of *Prunus*. *Science* 131:825-826.

Brown, L.C., J.C. Crane, and J.A. Beutel (1968). Gibberellic acid reduces cling peach flower buds. *California Agriculture* 22:7-8.

Burkholder, C.L. and M. McCown (1941). Effect of scoring and of α-Naphthyl acetic acid and amide spray upon fruit set and of the spray upon preharvest fruit drop. *Proceedings of the American Society for Horticultural Science* 38:117-120.

Byers, R.E. (1987). Peach bloom thinning with desiccating chemicals. *New York State Horticultural Society Newsletter*, April, pp. 1-8.

Byers, R.E., D.H. Carbaugh, and C.N. Presley (1990). The influence of bloom thinning and GA₃ sprays on flower bud numbers and distribution in peach trees. *Journal of Horticultural Science* 65:143-150.

Byers, R.E. and C.G. Lyons (1982). Flower bud removal with surfactant for peach thinning. *HortScience* 17:377-378.

Byers, R.E. and C.G. Lyons (1984). Flower thinning of peach with desiccating chemicals. *HortScience* 19:545-546.

Byers, R.E. and C.G. Lyons (1985). Peach flower thinning and possible sites of action of desiccating chemicals. *Journal of the American Society for Horticultural Science* 110:662-667.

Byers, R.E., C.G. Lyons, and R.L. Horsburgh (1982). Comparisons of Sevin and Vydate for thinning apples. *HortScience* 17:777-778.

Byers, R.E. and R.P. Marini (1994). Influence of blossom and fruit thinning on peach flower bud tolerance to an early spring freeze. *HortScience* 29:146-148.

Call, R.E. and S.D. Seeley (1989). Flower bud coating of spray oils delay dehardening and bloom in peach trees. *HortScience* 24:914-915.

Chan, B.C. and J.C. Cain (1967). The effect of seed formation on subsequent flowering in apple. *Proceedings of the American Society for Horticultural Science* 91:63-68.

Chandler, W.H. (1907). The winter killing of peach buds as influenced by previous treatment. *Missouri Agricultural Experiment Station Bulletin* 74.

Childers, N.F., J.R. Morris, and G.S. Sibbett (1995). *Modern Fruit Science*, Tenth Edition. Gainesville, FL: Horticulture Publications.

Clanet, H. and Y.C. Salles (1976). Etude de l'action de l'acide gibbérellique sur le développement des ébauches florales chez le pêcher: Conséquences pratiques. *Annales de l'Amelioration des Plantes* 26:285-294.

Corgan, J.N. and F.B. Widmoyer (1971). The effects of gibberellic acid on flower differentiation, date of bloom, and flower hardiness of peach. *Journal of the American Society for Horticultural Science* 96:54-57.

Costa, G., G. Vizzotto, C. Malossini, and A. Ramina (1995). Biological activity of a new chemical agent for peach flower thinning. *Acta Horticulturae* 394:123-125.

Coston, D.C. (1986). Effects of paclobutrazol on peaches. *Acta Horticulturae* 179:575.

Crowe, A.D. (1965). Effect of thinning sprays on metabolism of growth substances in the apple. *Proceedings of the American Society for Horticultural Science* 86:23-27.

Curry, E.A. and M.W. Williams (1991). Environmental factors in blossom thinning apple trees with monocarbamide dihydrogensulfate (MCDS). *Proceedings of the Plant Growth Regulator Society of America 18th Annual Meeting*, p. 168.

Denne, M.P. (1960). The growth of apple fruitlets and the effect of early thinning on fruit development. *Annals of Botany* 24:397-406.

Dennis, F.G., L.J. Edgerton, and K.G. Parker (1970). Effects of gibberellin and alar sprays upon fruit set, seed development, and flowering of 'Bartlett' pear. *HortScience* 5:158-160.

Deyton, D.E., C.E. Sams, and J.C. Cummins (1992). Application of dormant oil to peach trees modifies bud-twig internal atmosphere. *HortScience* 27:1304-1305.

Di Marco, L., T. Caruso, F.P. Marra, and A. Motisi (1992). Research on flower thinning of early-ripening peach and nectarine with urea. *Fruit Varieties Journal* 46:186-190.

Dorsey, M.J. (1935). Nodal development of the peach shoot as related to fruit bud formation. *Proceedings of the American Society for Horticultural Science* 33:245-247.

Dorsey, M.J. and R.L. McMunn (1927). Relation of the time of thinning peaches to the growth of fruit and tree. *Proceedings of the American Society for Horticultural Science* 24:221-228.

Edgerton, L.J. (1948). Peach fruit bud hardiness as affected by blossom thinning treatments. *Proceedings of the American Society for Horticultural Science* 52:112-114.

Edgerton, L.J. (1966). Some effects of gibberellin and growth retardants on bud development and cold hardiness of peach. *Proceedings of the American Society for Horticultural Science* 88:197-203.

Edgerton, L.J. and W.J. Greenhalgh (1969). Regulation of growth, flowering and fruit abscission with 2-chloroethanephosphonic acid. *Journal of the American Society for Horticultural Science* 94:11-13.

Elving, D.C. and R.A. Cline (1993). Benzyladenine and other chemicals for thinning 'Empire' apple trees. *Journal of the American Society for Horticultural Science* 118:593-598.

Facteau, T.J., K.E. Rowe, and N.E. Chestnut (1985). Firmness of sweet cherry fruit following multiple applications of gibberellic acid. *Journal of the American Society for Horticultural Science* 110:775-777.

Facteau, T.J., K.E. Rowe, and N.E. Chestnut (1989). Flowering in sweet cherry in response to application of gibberellic acid. *Scientia Horticulturae* 38:239-245.

Fallahi, E. (1997). Applications of endothalic acid, pelargonic acid, and hydrogen cyanamide for blossom thinning in apple and peach. *HortTechnology* 7:395-399.

Fallahi, E., B.R. Simons, J.K. Fellman, and W.M. Colt (1992). Use of hydrogen cyanamide for apple and plum thinning. *Plant Growth Regulation* 11:435-439.

Farley, A.J. (1923). Factors that influence the effectiveness of peach thinning. *Proceedings of the American Society for Horticultural Science* 20:145-151.

Faust, M. (1989). *Physiology of Temperate Zone Fruit Trees*. New York: Wiley.

Fitch, L.B., D.E. Ramos, and J.T. Yeager (1972). Tree shaker thinning of French prune. *California Agriculture* 26:5-6.

Forshey, C.G. (1987). A review of chemical fruit thinning. *New England Fruit Meeting Proceedings, Annual Meeting of the Massachusetts Fruit Grower Association,* North Amherst, Massachusetts, 93:68-73.

Fulford, R.M. (1966). The morphogenesis of apple buds. III. The inception of flowers. *Annals of Botany* 30:207-219.

Glenn, D.M., D.L. Peterson, D. Giovannini, and M. Faust (1994). Mechanical thinning of peaches is effective postbloom. *HortScience* 29:850-853.

Goffinet, M.C., T.L. Robinson, and A.N. Lakso (1995). A comparison of 'Empire' apple fruit size and anatomy in unthinned and hand-thinned trees. *Journal of Horticultural Science* 70:375-387.

Greene, D.M. (1981). Growth regulator application and cultural techniques to promote early fruiting of apples. In *Tree Fruit Growth Regulators and Chemical Thinning* (The Proceedings of the 1981 Pacific Northwest Tree Fruit Shortcourse), R.B. Tukey and M.W. Williams (Eds.). Pullman,WA: Washington State University, Cooperative Extension, pp.117-146.

Greene, D.W. (1995). Thidiazuron effects on fruit set, fruit quality, and return bloom of apple. *HortScience* 30:1238-1240.

Greene, D.W. and W.R. Autio (1989). Evaluation of benzyladenine as a chemical thinner on 'McIntosh' apples. *Journal of the American Society for Horticultural Science* 114:68-73.

Greene, D.W., W.R. Autio, J.A. Erf, and Z.Y. Mao (1992). Mode of action of benzyladenine when used as a chemical thinner on apples. *Journal of the American Society for Horticultural Science* 117:775-779.

Griggs, W.H. and B.T. Iwakiri (1961). Effects of gibberellin and 2,4,5-trichlorophenoxyproprionic acid sprays on 'Bartlett' pear trees. *Proceedings of the American Society for Horticultural Science* 77:73-89.

Gu, S. and P.B. Lombard (1994). Effect of crop density on vegetative growth, flowering, fruit size, and yield in 'Nijisseiki' pear trees. *Acta Horticulturae* 367:271-277.

Gur, A., E. Harcabi, and A. Breuer-Mizrahi (1993). Control of peach flowering with gibberellins. *Acta Horticulturae* 329:183-186.

Hall, F.R., R.A. Downer, J.A. Cooper, T.A. Ebert, and D.C. Ferree (1997). Changes in spray retention by apple leaves during a growing season. *HortScience* 32:858-860.

Harley, C.P., J.R. Magness, M.P. Mosure, L.A. Fletcher, and E.S. Degman (1942). Investigations on the cause and control of biennial bearing of apple trees. *USDA Technical Bulletin 792* (March).

Harrold, T.J. (1935). Comparative study of the developing and aborting fruits of *Prunus persica*. *Botanical Gazette* 96:505-520.

Hartmann, W. (1984). Effect of growth regulators on fruit, seed, and vegetative development of self-sterile plum cultivars (in German). *Gartenbauwissenschaft* 49:162-169.

Havis, A.L. (1962). Effects of time of fruit thinning of 'Redhaven' peach. *Journal of Horticultural Science* 80:172-176.

Hibbard, A.D. and A.E. Murneek (1950). Thinning peaches with hormone sprays. *Proceedings of the American Society for Horticultural Science* 56:65-69.

Hoad, G.V. (1978). The role of seed derived hormones in the control of flowering in apple. *Acta Horticulturae* 80:93-103.

Hull, J. and L.N. Lewis (1959). Response of one-year-old cherry and mature bearing cherry, peach, and apple trees to gibberellin. *Proceedings of the American Society for Horticultural Science* 74:93-100.

Johnson, D.S. (1994). Influence of time of flower and fruit thinning on the firmness of 'Cox's Orange Pippin' apples at harvest and after storage. *Journal of Horticultural Science* 69:197-203.

Jones, K.M., B. Graham, S.A. Bound, and M.J. Oakford (1993). Preliminary trials to examine the effects of ethephon as a thinner of 'Gala' and 'Jonagold' apples. *Journal of Horticultural Science* 68:139-147.

Jones, K.M., T.B. Koen, M.J. Oakford, and S.A. Bound (1989). Thinning 'Red Fuji' apples with ethephon or NAA. *Journal of Horticultural Science* 64:527-535.

Jones, K.M., T.B. Koen, M.J. Oakford, and S.A. Bound (1990). Thinning 'Red Fuji' apples using ethephon at two timings. *Journal of Horticultural Science* 65: 381-384.

Klein, J.D. and S. Cohen (1995). High concentrations of monocarbamide dihydrogensulfate are needed to thin nectarine blooms in Israel. *HortScience* 30:628.

Knight, J.N. (1978). Chemical thinning of the apple cultivar 'Laxton's Superb'. *Journal of Horticultural Science* 53:63-66.

Knight, J.N. (1980). Regulation of cropping and fruit quality of 'Conference' pears by the use of gibberellic acid and thinning. I. The effect of hand thinning of blossom in conjunction with gibberellic acid application. *Journal of Horticultural Science* 55:33-39.

Knight, J.N. (1986). Fruit thinning with carbaryl. *Acta Horticulturae* 179:707-708.

Knight, J.N. and G. Browning (1986). Regulation of 'Conference' pear cropping with gibberellic acid and ethephon or paclobutrazol. *Acta Horticulturae* 179:337-342.

Knight, J.N. and J.E. Spencer (1987). Timing of application of carbaryl used as an apple fruitlet thinner. *Journal of Horticultural Science* 62:11-16.

Li, S.H., C. Bussi, J. Hugard, and H. Clanet (1989). Critical period of flower bud induction in peach trees associated with shoot length and bud position. *Gartenbauwissenshaft* 54:49-53.

Li, S.H., Z.Q. Meng, T.H. Li, H.Z. Liu, and Y.C. Tu (1995). Critical period of flower bud induction in 'Red Fuji' and 'Ralls Janet' apple trees. *Gartenbauwissenshaft* 60:240-245.

Lichou, J., M. Jay, L. Gonsolin, M.L. Massacrier, and G. Du Fretay (1997). Armothin: A new chemical agent efficient for peach blossom thinning. *Acta Horticulturae* 451:683-689.

Lichou, J., M. Jay, and M.C. Prost (1994). Chemical thinning on peach tree with Armothin. *Seminaire AKZO-NOBEL Stellenbosch,* January 1994.

Lichou, J., M. Jay, and M.C. Prost (1995). L'éclaircissage chimique du Pêcher: Nouvelles perspectives. *INFOS-Ctifl* No. 110:28-33.

Lilleland, O. (1950). Effects of blossom spray thinning on size and yield of prunes. *California Fruit and Grape Grower*, March.

Looney, N.E., M. Beulah, and K. Yokota (1998). Chemical thinning of 'Fuji' apple. *Compact Fruit Tree* 31:55-57.

Luckwill, L.C. (1970). The control of growth and fruitfulness of apple trees. In *Physiology of Tree Crops,* L.C. Luckwill and C.V. Cutting (Eds.). London: Academic Press, pp. 237-254.

Luckwill, L.C. (1974). A new look at the process of fruit bud formation in apple. *Proceedings of the XIX International Horticultural Congress*, Warsaw, 3:235-245.

Luckwill, L.C. and J.M. Silva (1979). The effects of daminozide and gibberellic acid on flower initiation, growth and fruiting of apple cv. Golden Delicious. *Journal of Horticultural Science* 54:217-223.

MacDaniels, L.H. and E.M. Hildebrand (1939). A study of pollen germination upon the stigmas of apple flowers treated with fungicides. *Proceedings of the American Society for Horticultural Science* 37:137-140.

Marini, R.P. (1996). Chemically thinning spur 'Delicious' apples with carbaryl, NAA, and ethephon at various stages of fruit development. *HortTechnology* 6:241-246.

Marini, R.P. (1997). Oxamyl is an effective apple fruit thinner when used alone or in combination with other thinners. *HortTechnology* 7:253-258.

Marsh, H.V., F.W. Southwick, and W.D. Weeks (1960). The influence of chemical thinners on fruit set and size, seed development, and preharvest drop of apples. *Proceedings of the American Society for Horticultural Science* 75:5-21.

Martin, G.C. (1973). Peach fruit-set and abscission. *Acta Horticulturae* 34:345-352.

Martin, G.C., L.B. Fitch, G.S. Sibbett, G.L. Carnill, and D.E. Ramos (1975). Thinning French prune (*Prunus domestica* L.) with (2-chloroethyl)phosphonic acid. *Journal of the American Society for Horticultural Science* 100:90-93.

McLaughlin, J.M. and D.W. Greene (1984). Effects of BA, GA_{4+7}, and daminozide on fruit set, fruit quality, vegetative growth, flower initiation, and flower quality of 'Golden Delicious' apple. *Journal of the American Society for Horticultural Science* 109:34-39.

McLaughlin, J.M. and D.W. Greene (1991). Fruit and hormones influence flowering of apple. I. Effect of cultivar. *Journal of the American Society for Horticultural Science* 116:446-449.

Miller, S.S. (1988). Plant bioregulators in apple and pear culture. *Horticultural Reviews* 10:309-401.

Murneek, A.E. and A.D. Hibbard (1947). Investigations on thinning of peaches by means of caustic and "hormone" sprays. *Proceedings of the American Society for Horticultural Science* 50:206-208.

Myers, R.E., D.E. Deyton, and C.E. Sams (1996). Applying soybean oil to dormant peach trees alters internal atmosphere, reduces respiration, delays bloom, and thins flower buds. *Journal of the American Society for Horticultural Science* 121:96-100.

Myers, R.H. (1982). Fruit thinning and russeting of oxamyl on apple. *HortScience* 17:658-659.

Myers, S.C., A. King, and A.T. Savelle (1993). Bloom thinning of 'Winblo' peach and 'Fantasia' nectarine with monocarbamide dihydrogensulfate. *HortScience* 28:616-617.

Olien, W.C., R.W. Miller, C.J. Graham, E.R. Taylor, and M.E. Hardin (1995). Effects of combined applications of ammonium thiosulphate and fungicides on fruit load and blossom blight and their phytotoxicity to peach trees. *Journal of Horticultural Science* 70:847-854.

Painter, J.W. and G.E. Stembridge (1972). Peach flowering response as related to time of gibberellin application. *HortScience* 7:389-390.

Proebsting, E.L. and H.H. Mills (1964). Gibberellin-induced hardiness responses in 'Elberta' peach flower buds. *Proceedings of the American Society for Horticultural Science* 85:134-140.

Quinlan, J.D. and A.D. Preston (1968). Effects of thinning blossom and fruitlets on growth and cropping of 'Sunset' apple. *Journal of Horticultural Science* 43:373-381.

Rosenberger, D.A., F.W. Meyer, and C.A. Engle (1994). Summer fungicides applied to 'Liberty' apple trees affect timing of autumn leaf drop and effectiveness of fruit thinning with NAA the next year. *Fruit Varieties Journal* 48:55-56.

Sharples, R.O. (1968). Fruit thinning effects on the development and storage quality of 'Cox's Orange Pippin' apple fruit. *Journal of Horticultural Science* 43:359-371.

Shoemaker, J.S. (1933). Certain advantages of early thinning of 'Elberta'. *Proceedings of the American Society for Horticultural Science* 30:223-224.

Sibbett, G.S. and G.C. Martin (1982). Cumulative effects of ethephon as a fruit thinner on French prune (*Prunus domestica* L.). *HortScience* 17:665-666.

Southwick, F.W. and W.D. Weeks (1949). Chemical thinning of apples at blossom time and up to four weeks from petal fall. *Proceedings of the American Society for Horticultural Science* 53:143-147.

Southwick, F.W., W.D. Weeks, E. Sawada, and J.F. Anderson (1962). The influence of chemical thinners and seeds on the growth rate of apples. *Proceedings of the American Society for Horticultural Science* 80:33-42.

Southwick, S.M. and R. Fritts (1995). Commercial chemical thinning of stone fruit in California by gibberellins to reduce flowering. *Acta Horticulturae* 394:135-147.

Southwick, S.M., K.G. Weis, and J.T. Yeager (1996). Bloom thinning 'Loadel' cling peach with a surfactant. *Journal of the American Society for Horticultural Science* 121:334-338.

Southwick, S.M., K.G. Weis, J.T. Yeager, J.K. Hasey, and M.E. Rupert (1998). Bloom thinning of 'Loadel' cling peach with a surfactant: Effects of concentration, carrier volume, and differential applications within the canopy. *HortTechnology* 8:55-58.

Southwick, S.M., K.G. Weis, J.T. Yeager, and H. Zhou (1995). Controlling cropping in 'Loadel' cling peach using gibberellin: Effects on flower density, fruit distribution, fruit firmness, fruit thinning, and yield. *Journal of the American Society for Horticultural Science* 120:1087-1095.

Southwick, S.M. and J.T. Yeager (1991). Effects of postharvest gibberellic acid on return bloom of 'Patterson' apricot. *Acta Horticulturae* 293:459-466.

Southwick, S.M. and J.T. Yeager (1995). Use of gibberellin formulations for improved fruit firmness and chemical thinning in 'Patterson' apricot. *Acta Horticulturae* 384:425-429.

Southwick, S.M., J.T. Yeager, and K.G. Weis (1997). Use of gibberellins on 'Patterson' apricot (*Prunus armeniaca*) to reduce hand thinning and improve fruit size and firmness: Effects over three seasons. *Journal of Horticultural Science* 72:645-652.

Southwick, S.M., J.T. Yeager, and H. Zhou (1995). Flowering and fruiting in 'Patterson' apricot (*Prunus armeniaca*) in response to postharvest application of gibberellic acid. *Scientia Horticulturae* 60:267-277.

Spencer, S. and G.A. Couvillon (1975). The relationship of node position to bloom date, fruit size, and endosperm development of the peach, *Prunus persica* (L.)

Batsch cv. Sullivan's Elberta. *Journal of the American Society for Horticultural Science* 100:242-244.

Stembridge, G.E. and C.E. Gambrell (1971). Thinning peaches with bloom and postbloom applications of 2-chloroethylphosphonic acid. *Journal of the American Society for Horticultural Science* 96:7-9.

Stembridge, G.E. and J.H. LaRue (1969). The effect of potassium gibberellate on flower bud development in the 'Redskin' peach. *Journal of the American Society for Horticultural Science* 94:494-495.

Stembridge, G.E. and G. Morrell (1972). Effect of gibberellins and 6-benzyladenine on the shape and fruit set of 'Delicious' apples. *Journal of the American Society for Horticultural Science* 97:464-467.

Stopar, M., B.L. Black, and M.J. Buckovac (1997). The effect of NAA and BA on carbon dioxide assimilation by shoot leaves of spur-type 'Delicious' and 'Empire' apple trees. *Journal of the American Society for Horticultural Science* 122:837-840.

Taylor, B.H. and D. Geisler-Taylor (1998). Flower bud thinning and winter survival of 'Redhaven' and 'Cresthaven' peach in response to GA_3 sprays. *Journal of the American Society for Horticultural Science* 123:500-508.

Tromp, J. (1982). Flower-bud formation in apple as affected by various gibberellins. *Journal of Horticultural Science* 57:277-282.

Tukey, H.B. (1936). Development of cherry and peach fruits as affected by destruction of the embryo. *Botanical Gazette* 98:1-24.

Tukey, H.B. and O. Einset (1938). Effect of fruit thinning on size, color, and yield of peaches and on growth and blossoming of the tree. *Proceedings of the American Society for Horticultural Science* 36:314-319.

Turner, J.N. (1973). Gibberellic acid for controlling fruit production of pears. *Acta Horticulturae* 34:287-297.

Unrath, C.R. (1974). The commercial implications of gibberellin A_4A_7 plus benzyladenine for improving shape and yield of 'Delicious' apples. *Journal of the American Society for Horticultural Science* 99:381-384.

Ward, D.L. (1993). Reducing flower bud density of 'Redkist' peach with GA_3. MS Thesis. Southern Illinois University, Carbondale, Illinois.

Way, D.W. (1967). Carbaryl as a fruit thinning agent. II. Concentration and time of application. *Journal of Horticultural Science* 42:355-365.

Webster, A.D. and L. Andrews (1985). Fruit thinning 'Victoria' plums (*Prunus domestica* L.): Preliminary studies with paclobutrazol. *Journal of Horticultural Science* 60:193-199.

Weinberger, J.H. (1941). Studies on time of peach thinning from blossoming to maturity. *Proceedings of the American Society for Horticultural Science* 38:137-140.

Wertheim, S.J. (1982). Fruit russeting in apple as affected by various gibberellins. *Journal of Horticultural Science* 57:283-288.

Westwood, M.N. and H.D. Billingsley (1967). Cell size, cell number, and fruit density of apples as related to fruit size, position in the cluster, and thinning method. *Proceedings of the American Society for Horticultural Science* 91:51-62.

Williams, M.W. (1979). Chemical thinning of apples. *Horticultural Reviews* 1:270-300.

Williams, M.W. (1993a). Sulfcarbamide, a blossom-thinning agent for apples. *HortTechnology* 3:322-324.

Williams, M.W. (1993b). Comparison of NAA and carbaryl petal-fall sprays on fruit set of apples. *HortTechnology* 3:428-429.

Williams, M.W. and L.P. Batjer (1964). Site and mode of action of 1-Naphthyl N-methylcarbamate (Sevin) in thinning apples. *Proceedings of the American Society for Horticultural Science* 85:1-10.

Williams, M.W., S.A. Bound, J. Hughes, and S. Tustin (1995). Endothall: A blossom thinner for apples. *HortTechnology* 5:257-259.

Williams, M.W. and L.J. Edgerton (1981). Chemical fruit thinning of apple and pear. *USDA Agricultural Information Bulletin* 289:1-22.

Wismer, P.T., J.T.A. Proctor, and D.C. Elving (1995). Benzyladenine affects cell division and cell size during fruit thinning. *Journal of the American Society for Horticultural Science* 120:802-807.

Yokota, K., K. Murashita, S. Takita, M. Nonaka, S. Kato, and T. Suyama (1995). Flower thinning effect of synthetic auxins on 'Fuji' apple. *Acta Horticulturae* 394:105-109.

Zilkah, S., I. Klein, and I. David (1988). Thinning peaches and nectarines with urea. *Journal of Horticultural Science* 63:209-216.

Index

Page numbers followed by the letter "i" indicate illustrations; those followed by the letter "t" indicate tables.

T - #0504 - 101024 - C0 - 212/152/15 - PB - 9781560228967 - Gloss Lamination